Chemistry Survival Skills

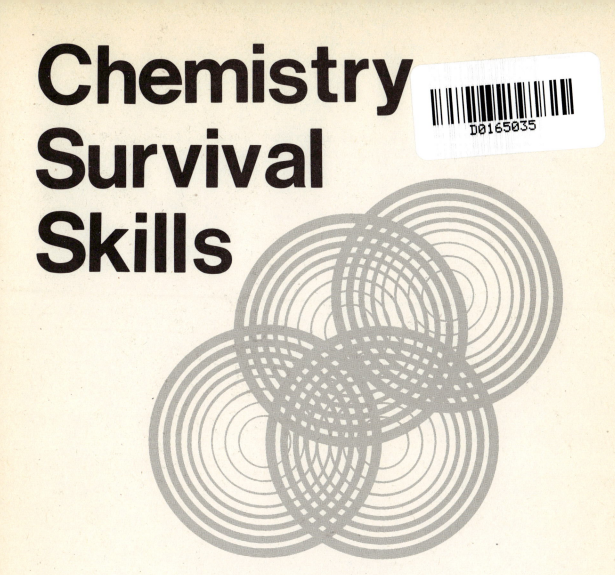

Margaret A. Brault
Margaret L. MacDevitt

Northern Michigan University

D. C. HEATH AND COMPANY
Lexington, Massachusetts   Toronto

Dedicated to my parents and to Mac, Shaun, and Mattie, whose patience, understanding, and humor continue to teach me the most valuable life survival skills.

M. L. M.

Published simultaneously in Canada.

Printed in the United States of America.

International Standard Book Number: 0-669-17143-3

INTRODUCTION

We have written *Chemistry Survival Skills* with the intention of helping chemistry students be more successful in their courses. Over the years, we have seen students struggle through chemistry, often failing or doing poorly because they lacked the necessary study skills.

The purpose of this book is to teach you skills to not only help you "survive" chemistry, but to learn more and to enjoy it. We do not promise that you will get an A in your chemistry class if you follow all of our recommendations, but it would surprise us if your chemistry grades did not improve. Sounds like a tall order to fill but we are confident that you will find the skills presented to be helpful.

You will discover that the book is applicable not only to chemistry but also to other courses. The skills taught in the concentration and test-taking chapters, for example, are also applicable to athletics, theatre, or any situation where performance is required. Most of the skills are based on successful teachings from general study skills books and programs. We have simply modified them to make them more useful for chemistry.

Chemistry Survival Skills is composed of seven chapters which focus on skill building in the following areas: time management, concentration, listening and note-taking, reading the chemistry text, problem solving, test-taking, and the laboratory. Over the years we have observed that successful chemistry students are highly skilled in these areas.

Acknowledgments

We would first of all like to thank the struggling students who inspired us to write this book.

At Northern Michigan University, we would like to express our deepest thanks to Sandra B. Salo, John W. MacDevitt, Gerald D. Jacobs, Gail D. Griffith, P. Anna Ellendman, Jenny Granger, and Perrin Fenske for their interest, ideas, and editorial comments. We would also like to thank the members of the Northern Michigan University Chemistry Department for their support.

Thanks also to Mary Le Quesne, our Editor, for her support and patience in seeing this manuscript through to final publication.

We are very grateful for the Curriculum Development Grant awarded to us by the Faculty Grants Committee of Northern Michigan University which enabled us to write this book.

Last, but not least, we would like to acknowledge the University of Utah Study Skills Program, the Northern Michigan University Counseling Center Paraprofessional Program, Walter Pauk, David B. Ellis, John Langan, Michael Weissberg and Fred Breme—our study skills references whose ideas we borrowed and adapted.

M. A. B.

M. L. M.

CONTENTS

TO THE STUDENT

Before you begin reading, keep in mind that this book will require you to *work* in order to master the skills presented. *Practice* is necessary in order to retain the skills. You will learn more if you answer the questions at the beginning of each chapter, make notes to yourself as you read, and do the brief exercises at the end of each chapter. Be sure to have a pen or pencil in hand while reading. We also recommend that you have your chemistry textbook and your lecture notes nearby as you work through the book, as we may ask you to refer to them.

The book is divided into seven chapters:

1. Time Management for Chemistry

2. Improving Your Concentration

3. Listening and Note-Taking

4. Reading Your Chemistry Textbook

5. Chemistry Problem Solving

6. Taking Chemistry Tests

7. The Chemistry Laboratory

Because each chapter is independent of the others, you may start with any one of the seven chapters. This is especially helpful if you have already identified an area that needs immediate attention. For example, if you know you need help with your laboratory skills, start with Chapter 7, "The Chemistry Laboratory." If you don't know where to begin, we suggest you start with Chapter 1.

Each chapter has a brief introduction followed by a problem check list, which is designed to help you determine whether or not your skills need improvement in a particular area (for example, time management). If they do not, you are instructed to go to the next chapter.

If your skills do need improvement, you are instructed to read the skill-building tips. Focus on those tips that are useful to you. As you read, note on a piece of paper anything you want to do differently to improve your skills.

You may be unaware that your skills need improvement in some areas. For that reason, we recommend that you complete the problem check lists for all of the chapters.

At the end of each chapter is a skills check list, which is a list of statements that will help you clarify exactly what changes you need to make to improve your skills in that area. As you go through this end-of-the-chapter skills check list, you may find that you need to review one or more of the tips. For your convenience, we have included the appropriate tip number and page reference next to each skill statement.

In each chapter, once you have determined all the ways you can improve your skills, you will be asked to prioritize the changes you want to make, selecting three behavior changes to improve your skills. We ask that you follow our recommendations of only trying to make three changes per chapter at one time, so that you don't become overwhelmed and discouraged. Our book is designed so that you can go through it again and again, noting your progress and making further changes to improve your skills.

IF OUR BOOK DOES NOT HELP YOU

If you are doing poorly in chemistry and our book does not help you, ask yourself the following:

1.   Am I taking chemistry because I want to?

2.   Am I going to college because I want to?

3.   Do I have the math and chemistry background I need for this chemistry course?

If you are unsure about your answers to any of the above three questions or if your answer is no, you should talk to your advisor or an academic or personal counselor about whether chemistry is the course for you.

1 TIME MANAGEMENT FOR CHEMISTRY

Why is time management so important to chemistry?

As you've already discovered or should have discovered, chemistry courses are time-consuming. You are required to attend lectures, laboratories, and sometimes, discussions. You then need additional time to prepare for lecture and laboratory, review your notes, study, do homework problems, prepare for quizzes and exams, read the laboratory experiment, do the prelaboratory assignment, and write up the previous week's laboratory experiment.

It's obvious that you need to know how to manage your time efficiently if you're going to survive chemistry and that's what we'll be helping you with in this chapter.

Let's first take a look at some time management problems that chemistry students frequently report.

Some of the material in this chapter has been taken from *Reading and Study Skills*, by John Langan, pp. 15–18 and 42. Copyright © 1978 by McGraw-Hill Book Company. Reprinted by permission of the publisher.

Time Management Problems

Read the following list of problem statements and place a check mark in the appropriate space to the left to indicate whether the statements are true for you.

YES   NO

_____ _____   1. I always seem to get tired when I begin studying chemistry.

_____ _____   2. I never have enough time to get all my chemistry homework done.

_____ _____   3. I usually cram for tests.

_____ _____   4. Studying chemistry takes up so much of my time that I have little time for anything else.

_____ _____   5. I'm frequently late to my chemistry labs or lectures.

_____ _____   6. I often don't get around to doing my chemistry homework.

_____ _____   7. I constantly feel overwhelmed by all the work that I have to do.

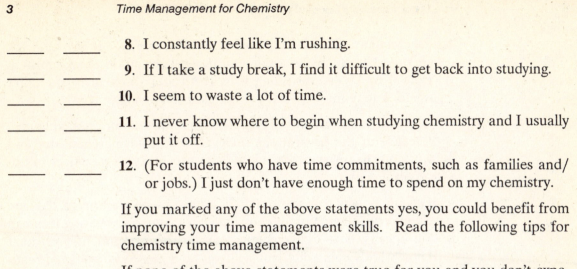

8. I constantly feel like I'm rushing.

9. If I take a study break, I find it difficult to get back into studying.

10. I seem to waste a lot of time.

11. I never know where to begin when studying chemistry and I usually put it off.

12. (For students who have time commitments, such as families and/or jobs.) I just don't have enough time to spend on my chemistry.

If you marked any of the above statements yes, you could benefit from improving your time management skills. Read the following tips for chemistry time management.

If none of the above statements were true for you and you don't experience any problems with time management, skip ahead to the next chapter, "Improving Your Concentration."

Time Management Tips

As you read the following tips, write down on a sheet of paper anything you want to do differently to improve your time management skills.

1. *Plan at least three hours of study time for each hour of lecture time.* In other words, if your class meets for three hours during the week, you should be spending nine hours outside preparing for class, cleaning up your notes, reading your text, working problems, and preparing for exams.

   For laboratories, allow one hour outside for every hour spent in the laboratory. If your chemistry lab meets for three hours, you should be spending three hours outside of lab, reading the laboratory experiment, doing the prelaboratory assignment, or writing up the previous week's experiment.

   You may need either additional or less study time than is indicated in the formulas above. This will depend on your background in chemistry and mathematics, your professor's expectations and requirements, and how well you want to do in the course.

2. *Schedule a regular time to study chemistry.* Having a regular study time is helpful for several reasons:

   a. This regularity will result in studying becoming a habit for you.

b. Procrastination takes a lot of time and energy. If studying for chemistry becomes an automatic thing that you do every morning from 9 to 11, like brushing your teeth after breakfast, you won't waste time telling yourself you should be studying.

You may also waste energy feeling guilty about not studying. Because of the frustrations involved with studying chemistry and working chemistry problems, it is an especially easy subject to put off until later.

c. Having a regular study time will allow you to stay up to date on chemistry assignments. You won't find yourself sweating it out in lab because you didn't get to your prelab assignment and don't know what's going on. Nor will you be trying to fit weeks of reading and problems into two nights before the test.

d. Research has demonstrated that a series of study sessions is more effective for learning material than is a single long cram session. This is especially true in chemistry, where you need time to digest and assimilate the seemingly endless list of equations, formulas, and concepts.

e. It's more fun to stay on top of things than to always be behind.

3. *Keep a daily, weekly, and semester schedule for yourself.*

   a. *Semester Schedule*: At the beginning of each semester take your syllabus for each class and write out all the assignments and long-term projects and their due dates.

   ■ Buy a schedule book or make up your own calendar for the term. (As long as it has the dates in it and lots of space to write on each date, you're fine.)

   ■ Write in all the assignments and project due dates that you have, and as you're given more assignments during the semester, add them to your schedule book.

   ■ Break the projects down into smaller chunks. If you have a term paper due at the end of the semester, you might start with the outline and when you'd like to have that completed; then your library work; and so on.

   ■ Write in all your regularly scheduled activities. Examples are your classes, laboratories, a regular work schedule, and regularly scheduled meetings. Include regularly scheduled study periods.

b. *Weekly Schedule*: Each week, fill in your semester schedule with activities planned for that week. Include activities such as meals, bedtime, study periods, and social commitments.

For your study periods, *be specific*. Write down exactly what you're going to work on during your chemistry study period, as opposed to just writing down "chemistry." For example, for Monday, 2:30–4:30 P.M. "Read Chapter 8, do 5 problems," or "Do prelab assignment for Tuesday lab." Being specific about what you're going to do will help you be more productive during your study period, as you won't be wasting time trying to decide what to do.

c. *Daily Schedule*: As a reminder for what you're going to do each day, write out each day's schedule on something like a 3 × 5 card that you can easily carry around with you as a reminder.

d. Allow 30 minutes each week to update your schedule and make revisions. Sunday evening is a good time for this.

4. *Plan study sessions of at least one hour*. If you schedule less than that, you may find that your study time is over with just as you're getting warmed up and involved with what you're doing.

   a. You can plan to study longer, but then be sure to take a 10-minute break after each 50-minute period to reenergize yourself.

   b. You may find that initially your attention span for chemistry is only 30 minutes, after which you may need a 10-minute break. That's fine. *You can increase the frequency of your breaks, but do not increase the time taken*. In other words, whether your study period is 30 or 50 minutes, don't take more than a 10-minute break.

5. *Reward yourself for studying*. Research indicates that people work better when they are rewarded for their efforts. Thus you might want to set up a reward system for yourself following productive study periods. Examples of rewards are a phone call to a friend or watching a favorite TV program.

   a. Remember to reward yourself *after* you study. The system doesn't work if you reward yourself before you study or if you don't study but reward yourself anyway.

   b. During your 10-minute breaks, do something to reenergize yourself. Do some exercises, get something to eat, walk around, etc.

6. *Schedule study periods before and after your chemistry lecture and laboratory*. By scheduling a study period

- before lecture, you will be able to read over the chapter which will be covered and review your notes from the previous lecture. In this way, the material covered that day will be fresh in your mind and more meaningful to you.

- after lecture, you can revise your notes while the lecture is still fresh for you.

- before laboratory, you will be able to read the experiment and do the prelaboratory assignment which will increase your understanding of the laboratory material and help you to be better organized for lab.

- after laboratory, you will have the time to do the needed calculations and complete the laboratory report while the experiment is fresh in your mind.

7. *Use your dead time*. If you have 10 to 15 minutes available, such as while you're making lunch or washing dishes, recite your chemistry to yourself or go over your 3 × 5 note cards (it may be helpful to always carry them with you for such times). (See Chapter 4, "Reading Your Chemistry Textbook," for explanation of 3 × 5 cards.)

  a. Going over material in your mind (recitation) facilitates learning and is one type of studying that doesn't require a distraction-free environment. Not only do you learn the material but you prove to yourself that you know it, which increases your satisfaction with yourself and decreases your anxiety.

  b. While reading your text and working the problems require a good chunk of time in a quiet study area, recitation can be helpful even if only done for a few minutes at a time.

8. *Schedule free time*. If you schedule some free time for yourself each day, you'll find that you'll use your study time more efficiently. During your free time, do things that you enjoy such as socializing, watching TV, going for a walk, or even daydreaming.

  a. If you're someone who doesn't schedule any free time for yourself, you may have noticed that you get burned out on studying and become less efficient. Your mind is likely to wander while you're trying to study or you may end up stealing free time during a scheduled study period. This results in your feeling guilty, so that you aren't able to enjoy your recreational time or get a break from your studies.

 b. Be sure to make time for recreation, but again, try to remember the principle "study first, play later." Plan your recreation time to make it more satisfying. That is, a weekend skiing with friends, a movie that you really want to see, a dance you want to attend, and so on.

9. *Study during your peak energy times.*

 a. Know when your high-energy and low-energy times are during the day.

 b. Schedule your chemistry study periods and study periods for other difficult subjects at times when you're most refreshed and alert. Use your low-energy times for easier tasks such as recitation and review, or as your free time.

 c. Quite often, a little exercise during low-energy periods will boost your energy and allow you to continue with your studying. Taking a nap or doing something sedentary like watching TV is likely to further decrease your energy.

10. *Keep your schedule flexible so that you can exchange time instead of losing it.* If something comes up where you're going to have to miss a regularly scheduled study period during the week, instead of missing it altogether, trade times so that the study period is rescheduled.

You may find that your schedule requires constant readjustments. In that case, simply revise it each week or daily if necessary.

11. *Get plenty of sleep, exercise, and proper nutrition.* If you're tired, your body is out of shape, and you've just eaten a heavy meal it's obvious that your mind won't be very alert. It's important that you schedule time to eat regular meals each day, that you get regular exercise, and that you have a regular sleep pattern where you get six to eight hours of sleep each night.

If you want to make better use of your time and be as alert as possible in chemistry, you have to take care of your body too.

Time Management Skills Check List

To help you clarify what time management changes you need to make, answer yes or no to the following statements. If necessary, review the appropriate tips for "no" statements.

YES   NO

_____ _____ 1. I study chemistry at least three hours outside of class for every hour in lecture. (See Tip 1, page 3.)

_____ _____ 2. For every hour in lab, I spend one hour outside on laboratory preparation and write-ups. (See Tip 1, page 3.)

_____ _____ 3. I schedule regular times to study chemistry every week. (See Tip 2, page 3.)

_____ _____ 4. I use a semester schedule, weekly schedule, and daily schedule. (See Tip 3, page 4.)

_____ _____   5. My chemistry study sessions are at least one hour long. (See Tip 4, page 5.)

_____ _____   6. After studying chemistry for 50 minutes, I take a 10-minute break. (See Tip 4, page 5.)

_____ _____   7. After a productive study session, I reward myself by _____. (See Tip 5, page 5.)

_____ _____   8. My chemistry study periods are scheduled both before and after lectures and laboratories. (See Tip 6, page 6.)

_____ _____   9. I use the time I spend waiting and doing menial chores for chemistry recitation. (See Tip 7, page 6.)

_____ _____   10. I schedule recreational and relaxation time for myself each week. (See Tip 8, page 6.)

_____ _____   11. I know when my high- and low-energy times are each day, and I develop my time schedule accordingly. (See Tip 9, page 7.)

_____ _____   12. If something comes up that interferes with my study time, I don't lose that time; I reschedule it by exchanging times. (See Tip 10, page 7.)

_____ _____   13. I take care of my body with plenty of sleep, exercise, and proper nutrition. (See Tip 11, page 7.)

Review the statements to which you answered no. In addition, review the time management suggestions you have already written on a separate sheet of paper. Then, do the following:

A. Prioritize these lists and select three behaviors that you can change that will help you to improve your chemistry time management.

B. Write these behavior changes in the space below.

C. Be specific about what you want to change, how you are going to make the change, and when the new behavior will be implemented.

D. See if these changes help you better manage the time you spend on chemistry.

   1.

   2.

   3.

Do not try to make too many time management changes at one time because this can be overwhelming and lead to frustration. Once you have implemented the above three time management behavior changes and have seen improvement in your time management skills, we suggest that you again review how you manage your time. Then go through steps A through D and watch for further improvement.

2 IMPROVING YOUR CONCENTRATION

Why is concentration so crucial when studying chemistry?

?

In order to do well in chemistry, it is extremely important that you be able to concentrate while studying. There are numerous terms, symbols, laws, and formulas that must be learned exactly. In addition, most of the material to be learned is presented in mathematical terms. Your work involves a good deal of accuracy in being able to do mathematical manipulations and problem solving.

As a result, no matter how smart you are, how much scientific potential you have, or how well your time is managed, if you aren't able to concentrate on your chemistry while studying and/or in lecture, you're going to have a difficult time.

Let's take a look at some problems students sometimes report while studying chemistry. Concentration problems during the lecture are covered in Chapter 3, "Listening and Note-taking During the Chemistry Lecture."

Frequent Concentration Problems for Chemistry Students

Read the following list of problem statements and place a check mark in the appropriate space to the left to indicate whether the statements are true for you.

YES   **NO**

_____   _____   **1.** My mind often wanders when I'm trying to study my chemistry.

_____   _____   **2.** I'm easily bothered by noise when I study.

_____   _____   **3.** I often get hungry while I'm studying.

_____   _____   **4.** I often have a hard time concentrating on my chemistry problems.

_____   _____   **5.** My studying is often disrupted because I can't find the materials I need to work on my chemistry (pencil, paper, calculator, etc.).

_____   _____   **6.** I often fall asleep while studying.

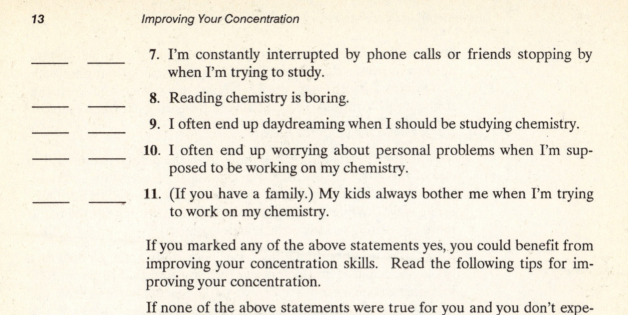

_____ _____   7. I'm constantly interrupted by phone calls or friends stopping by when I'm trying to study.

_____ _____   8. Reading chemistry is boring.

_____ _____   9. I often end up daydreaming when I should be studying chemistry.

_____ _____   10. I often end up worrying about personal problems when I'm supposed to be working on my chemistry.

_____ _____   11. (If you have a family.) My kids always bother me when I'm trying to work on my chemistry.

If you marked any of the above statements yes, you could benefit from improving your concentration skills. Read the following tips for improving your concentration.

If none of the above statements were true for you and you don't experience any problems with concentration while studying chemistry, skip ahead to the next chapter, "Listening and Note-taking."

Tips for Improved Concentration While Studying Chemistry

As you read the following tips, write down on a sheet of paper anything you want to do differently to improve your concentration skills while studying chemistry.

We'll discuss our Concentration Tips under two sections, "Eliminating External Distractions," and "Dealing with Internal Distractions," both of which interfere with studying.

External distractions are those distractions from your surroundings that make it difficult to study. Examples are TV, the stereo, a party next door, the refrigerator, and the telephone.

Internal distractions are those distractions that come from inside you that make it difficult to study. Examples are sleepiness, daydreaming, indecision, and hunger.

ELIMINATING EXTERNAL DISTRACTIONS

1. *Find a study area that is free of distractions.* If the telephone is ringing, friends are dropping by, and the stereo or TV are on, don't count on being able to focus on your chemistry. Find an area to study where you don't have to worry about interruptions.

a. Study in the library or another quiet area if at all possible. If you need to or prefer studying in your room, try closing your door with a "Do Not Disturb" sign on it. Unplug the phone and keep the stereo and TV turned off. You may find wearing ear plugs helpful.

b. *If studying in the library or a study room*:

- Use an individual study carrel away from the action as opposed to one of the large tables located in the midst of everything. Try to face away from traffic and don't sit near a window with an interesting view. (While studying, you don't want anything nearby that's more interesting to look at than your chemistry.)

- Don't look up when someone passes your study area. (You'll then get in the habit of not looking up.) If you find that to be a problem, make a mark on a piece of paper each time you look up. Monitoring undesirable behavior will help to decrease its frequency.

- If people come over to talk with you, quickly inform them that you are studying and that you will contact them later.

c. *Interruptions at times are unavoidable*. When interrupted, have a basic ritual you go through to resume studying. Examples are doing a few jumping jacks, straightening the papers on your desk, going to the bathroom, and getting a drink of water. Just make it something you do that you've come to associate with "I'm ready to focus on my studies."

d. *If you have children* who bother you while you study, try the following: When it's time to study, help the children get involved in an activity in anther room. Close your door and hang something on the door that signals them that you're studying and should not be disturbed, such as a hat or a red piece of paper. Remember that even pigeons can learn to discriminate; your children can too.

Tell your children that if they leave you alone to study for one hour, you'll reward them in some way. (Be sure to follow through or they won't believe you the next time.)

2. *Use your study area for studying only*. Don't use it as a place to chat, eat, or sleep. When you sit down in your study area, your main associations will be with studying. This will make it easier for you to focus on your studies when sitting there.

3. *Your study area should be well lit.*

   a. Working in a poorly lit study area will strain your eyes and make it difficult for you to concentrate. This will result in your becoming sleepy and inattentive.

   b. Make sure there are no flickering lights and eliminate any glare. To eliminate glare, use indirect lighting, have a good shade in your study lamp, and use a light-colored desk blotter.

4. *Make your study area a comfortable working place.*

   a. You should use an uncluttered table or desk that is large enough to spread out your materials. This will allow you to easily locate things.

   b. Have all of your materials handy when you begin studying, such as textbook, class notes, reference books, paper, pencil, calculator, clock, and calendar. That will cut down on interruptions to find things that you need.

   c. Your surroundings should be pleasant to you. The temperature should be comfortable and the ventilation good. Use a comfortable chair.

DEALING WITH INTERNAL DISTRACTIONS

5. *Schedule your study time.* This will reduce indecision about what to study and when. Spending your study hour trying to decide such things as whether you should prepare for lab or do some problems is a waste of time. If you find that this is often a distractor for you, review the time management chapter.

6. *Eliminate daydreaming while you're studying.* Don't try to get rid of all daydreaming in your life. It's a creative type of thinking that most humans engage in and it can be valuable and rewarding. However, it is important to find an appropriate time to daydream. And that appropriate time should definitely not be while you are studying chemistry. There are four things you can do to eliminate daydreaming while studying.

   a. As mentioned in the time management chapter, put a mark on a piece of paper every time you begin to daydream while you're studying. Monitoring undesirable behavior will decrease its frequency.

b. Just as you schedule your study time, also schedule time to daydream during the day. Use that daydreaming time as a reward when you have finished all your studying.

c. When it is time to daydream, get up and get out of your study space to do it. This will help you develop environmental cues for studying which are different from those associated with daydreaming.

d. Get actively involved in your reading or in working on your chemistry problems. Your mind can't think of two things at once, and so if it's focused on studying it won't be able to daydream at the same time.

e. When your mind begins to wander, don't criticize yourself. That will interfere with your studying more than mind-wandering does. Simply use one or all of the above steps to bring yourself back to your chemistry.

7. *Don't let worries about personal problems interfere with your studying*. As with daydreaming, you can't simply stop worrying about your personal problems, but you can take some positive steps so that those worries don't interfere with your studying.

    a. Set aside some time to deal with your personal problems either by yourself or with the help of a friend or counselor. Develop a plan of action to reduce your concern about your studies.

    b. If, while studying, one of your worries keeps bothering you, note it on a separate piece of paper and set aside a time to think about it some more.

    c. If you are a hard core worrier, you may need to schedule 20–30 minutes of worry-time for yourself each day. Don't do anything else during that time except worry and then don't allow yourself to worry at any other time during the day.

8. *Eat a well-balanced diet*. Eating a heavy meal will drag you down and make you tired; not eating at all results in hunger; both conditions will interfere with your ability to concentrate on chemistry. By eating healthfully and regularly throughout the day, you'll be more alert and better able to focus your attention on your studies.

9. *Be careful with coffee, alcohol, and drugs*.

    a. While the caffeine in coffee can work well for waking you up to study, too much can make you jittery and unable to concentrate. Caffeine also interferes with sleep, which may cause concentration difficulties.

    In addition, what goes up always comes down, and the same is true with the effects of caffeine. If you get lots of energy from coffee, expect in a few hours to feel depleted. Watch out for anxiety and depression which can also result from too much coffee.

    b. It's obvious that drinking alcohol and taking drugs will interfere with your ability to concentrate on your studies and to retain information, especially if you're hung over. If you are going to imbibe, do it *after* studying and don't let it disrupt your studies.

10. *Exercise regularly*. Find a form of exercise that suits you and that you can do regularly.

    a. Regular exercise that improves cardiovascular fitness (for example, aerobics, jogging, or swimming) makes you more alert. You'll need less sleep, and will feel better in general. It's often difficult to start exercising, but once you do, you'll notice positive changes.

b. Don't over-exercise, as that can result in injuries and fatigue.

11. *Eliminate mental fatigue while you're studying chemistry*. Mental fatigue is usually simple boredom. Here are things you can do to reduce boredom:

    a. As mentioned in the time management chapter, know your high- and low-energy times. Schedule your study periods for chemistry and other difficult subjects during those high-energy times when you're most alert.

    b. Divide your studying time among different subjects. Begin with chemistry the first hour if that's your most difficult subject. Reward yourself by spending your second hour on something that's easier.

    c. Be sure to take study breaks (10 minutes per hour) and try to get up and move around during those 10 minutes.

    d. Get *actively involved in studying chemistry* by doing the following:

    - Have a pencil, paper, and calculator nearby and use them as you study.

    - Underline in your text.

    - Form a study group. Make sure the other students are serious about chemistry or you may simply end up with another social group.

    - Check out other introductory chemistry texts from the library to get some different perspectives and explanations of the material. This will also give you additional problems to work for added review.

12. *Don't get discouraged*. Having difficulties with chemistry is not the end of the world. If you are too serious about it, you'll be tense and have an even more difficult time concentrating. Try to be relaxed when you're studying. When you don't understand something, instead of screaming with frustration, think of it as a challenge and do the best you can.

Even though your chemistry professor does the problems effortlessly while you may have difficulties knowing where to start, keep in mind that the professor started out in an introductory course too, and probably experienced the same types of problems and frustrations you are experiencing. Look at what persistence does! You can do it too!

13. *Remember the relationship between time management and concentration*. If you improve your concentration skills but aren't managing your time efficiently, you won't have time to use them. Likewise, if your time management skills are good and you have plenty of time to study, but are unable to concentrate while studying, those time management skills aren't going to work for you either.

Concentration Skills Check List

To help you clarify what changes you need to make to improve your concentration on chemistry, answer yes or no to the following statements. If necessary, review the appropriate tips for "no" statements.

YES   NO

_____   _____   1. My study area is distraction free. (See Tip 1, page 13.)

_____   _____   2. I have designed my study area and scheduled my study periods so that it is unlikely that I will be interrupted while studying. (See Tip 1, page 13.)

_____   _____   3. When interrupted, I have a ritual I go through to help me get back into studying. (See Tip 1, page 13.)

_____   _____   4. My study area is for studying only. (See Tip 2, page 14.)

_____   _____   5. The lighting in my study area is good. (See Tip 3, page 15.)

_____   _____   6. My study area is comfortable. (See Tip 4, page 15.)

_____   _____   7. My study area is well organized. (See Tip 4, page 15.)

_____   _____   8. I schedule my chemistry study times. (See Tip 5, page 15.)

_____   _____   9. It is true that daydreaming *does not* interfere with my studying chemistry. (See Tip 6, page 15.)

_____   _____   10. It is true that worries about personal problems *do not* interfere with my concentration on chemistry. (Tip 7, page 17.)

_____   _____   11. I have a healthy diet. (See Tip 8, page 17.)

_____   _____   12. I am careful not to abuse coffee, alcohol, or drugs. (See Tip 9, page 17.)

_____ _____ **13**. I exercise regularly. (See Tip 10, page 17.)

_____ _____ **14**. When I get tired while studying chemistry, I know how to quickly reenergize. (See Tip 11, page 18.)

_____ _____ **15**. I maintain my confidence while studying chemistry, even when I'm frustrated and don't understand the material. (See Tip 12, page 18.)

_____ _____ **16**. I use time management principles. (See Tip 13, page 19.)

Review the statements to which you answered no. In addition, review the suggestions for concentration improvement you have already written on a separate sheet of paper. Then, do the following:

A. Prioritize these lists and select three behaviors that you can change that will help you to improve your concentration on chemistry.

B. Write these behavior changes in the space below.

C. Be specific about what you want to change, how you are going to make the change, and when the new behavior will be implemented.

D. See if these changes help you better concentrate on chemistry.

    1.

    2.

    3.

Do not try to make too many changes at one time because this can be overwhelming and lead to frustration. Once you have implemented the above three concentration behavior changes and have seen improvement in your concentration skills, we suggest that you again review your concentration habits. Then go through steps A through D and watch for further improvement.

3 LISTENING AND NOTE-TAKING

What makes listening
and note-taking
during chemistry
lectures different?

1. Students are often intimidated by chemistry. They feel overwhelmed by the language, numbers, mathematics, formulas, and equations. In many ways chemistry is like a foreign language and, until you understand the meanings of the terms, listening to a lecture can be overwhelming and intimidating.

For example, every chemistry student encounters the term *mole* very early in the course. To the uninformed student, it could be a blemish on the skin, an insectivorous mammal, or an undercover agent. However, none of the previous definitions agrees with the chemical definition of the word mole. In order to be successful in chemistry, you must learn that, chemically speaking, a mole is 6.02×10^{23} molecules, atoms, or other species. There are numerous terms like this in chemistry and unless you are familiar with them, you will have a difficult time listening to and understanding chemistry lectures.

Some of the material in this chapter has been taken from *Reading and Study Skills*, by John Langan, pp. 15–18 and 42. Copyright © 1978 by McGraw-Hill Book Company. Reprinted by permission of the publisher.

2. Chemistry professors often use mathematics (effortlessly and flaw-lessly) as a means of communication during lectures. Students often find the extensive use of mathematics in chemistry frustrating for many reasons. A story problem is written on the board, interpreted, analyzed, and solved in what seems to be a matter of seconds by the professor. If you are unfamiliar with the mathematical manipulations you can easily become overwhelmed, frustrated, and anxious while lis-tening to the lecture. For others of you, this effortless and flawless performance by the professor may cause you to think that "any dumbo can do that," or, "chemistry is a piece of cake." Either way the result can be that you stop listening and your mind begins to wander.

3. Chemistry lectures involve a lot of note-taking. A tremendous amount of material is written on the board. In addition, verbal explanations are given for certain phenomena and concepts. You may become overwhelmed by the vast amount of material and think you need to use both hands in order to write everything down.

4. The contents of a chemistry lecture seem to be cut and dried and peppered with cold, hard facts. In the lower division courses, you will never read the original works of Einstein or Dalton, only a distillation of the main points of their theories. Unlike a literature course, there is no room (or very little at best) for individual interpretation of the concepts. You may find this barrage of abstract ideas hard to focus on, or more frankly, boring, which can make it difficult to concentrate during a lecture.

In this chapter, we'll be giving some tips on how to improve your listening and note-taking skills during chemistry lectures. But first, let's take a look at some problems chemistry students sometimes report they experience while listening and taking notes during lecture.

Problems Students Have When Trying to Listen and Take Notes During a Chemistry Lecture

Read the following list of problem statements and place a check mark on the appropriate space to the left to indicate whether the statements are true.

YES	NO	
____	____	1. I often find it difficult to pay attention during lecture.
____	____	2. I have trouble hearing the professor.
____	____	3. I often can't follow the lecture.
____	____	4. I am afraid to ask questions in class.
____	____	5. I believe that the professor thinks that I am stupid.
____	____	6. I sometimes run out of paper during the lecture.
____	____	7. I can't seem to write fast enough to get the entire lecture in my notes.
____	____	8. I feel overwhelmed by the amount of material covered during lecture.
____	____	9. I hate chemistry lectures.
____	____	10. I become sleepy during lecture.
____	____	11. I become anxious when the professor runs into overtime.
____	____	12. I often have knots in my stomach during chemistry lecture.

_____ _____   **13.** I can't figure out my notes after lecture.

_____ _____   **14.** I never feel that I understand the material completely.

_____ _____   **15.** I can't seem to remember the chemical symbols.

_____ _____   **16.** I usually have a number of mistakes in my notes.

_____ _____   **17.** I feel stupid when I don't understand what the professor is talking about.

If you marked any of the above statements yes, you could benefit from improving your listening and note-taking skills. Read the following tips for active listening and tips for note-taking during your chemistry lecture.

If none of the above statements were true for you and you don't experience any problems with listening and note-taking during a chemistry lecture, skip ahead to the next chapter, "Reading Your Chemistry Textbook."

Tips for Active Listening During a Chemistry Lecture

As you read the tips for active listening and note-taking during a chemistry lecture, keep in mind the problems you experience while listening and note-taking. Look for ways to change your behavior during these times in order to gain more from the lectures.

As you read the following, write down on a sheet of paper anything that you want to do differently to help you be a better listener.

1. *Attend all classes.* Lectures are a way for a professor to communicate the important material to you. Copying someone else's notes isn't a good idea for several reasons.

    a. Another student may make a mistake in his or her notes. For example, students often put incorrect chemical formulas in their notes (N_3H instead of the correct NH_3).

    b. Students often get behind in their note-taking and are unable to write complete solutions to problems in their notes. This can make studying from these notes more difficult.

    c. You will better understand and remember the important points if you hear the lecture yourself because you will know what the professor emphasized. The material will be more meaningful to you.

2. *Get plenty of sleep* and keep away from alcohol and drugs the night before lecture and lab. (That goes for the day of your lecture and lab too!)

3. *Read the assigned material* and review your notes from the last lecture before class. This helps you to listen better in lecture because you will be somewhat familiar with the terms and concepts before the lecture begins.

4. *Get to lecture early* and *sit as close to the front* of the room as you can. This eliminates distractions and helps you to better focus on the lecture. Common distractions that occur in the back of the classroom are talking, flirting, and eating.

5. *Have your pen and paper ready*, assume a good posture, and be ready to write when the professor begins to lecture.

6. If you're current with your homework assignments and reading, and still have trouble following the lecture, *ask questions* or request further information or examples for material that seems unclear.

   a. Most professors enjoy interacting with students during lecture because it breaks the monotony of talking to themselves.

   b. This gives the professor feedback so he or she knows which concepts and problems are difficult to understand.

   c. You can be sure that if you have a question there are other students who have the same question.

   d. In most cases, questions reflect positively on the student because they indicate an interest in the subject.

7. Be prepared to focus your attention on the professor and the lecture, not on your classmates, your personal problems, or the clock.

8. If your mind begins to wander during the lecture, make a mark each time it wanders. Research has shown that simply monitoring undesirable behavior makes it decrease in frequency. You can use each mark as a signal to focus your attention on your professor by changing your body position and getting yourself involved.

9. Remember this quote from Dante, "He listens well, who takes notes." *Taking notes during chemistry lectures helps you focus* on the material being presented. Your mind won't wander because you'll be more involved with the lecture.

No matter how smart you are, you're human and humans forget most of what they hear. Studies have shown that within two weeks you are likely to have forgotten 80 percent or more of what you have heard, and in four weeks you are lucky if 5 percent remains! TAKE NOTES! (For better note-taking skills, see the next section on note-taking.)

Listening Skills Check List

To help you clarify what listening changes you need to make, answer yes or no to the following statements. If necessary, review the appropriate tips for the "no" statements.

YES   NO

_____ _____   **1.** I attend all lectures. (See Tip 1, page 24.)

_____ _____   **2.** I rarely copy someone else's notes. (See Tip 1, page 24.)

_____ _____   **3.** I get plenty of sleep the night before lecture. (See Tip 2, page 25.)

_____ _____   **4.** I stay away from alcohol and drugs the night before class and the day of class. (See Tip 2, page 25.)

_____ _____   **5.** I read the assigned material before class. (See Tip 3, page 25.)

_____ _____   **6.** I get to lecture early. (See Tip 4, page 25.)

_____ _____   7. I sit close to the front of the lecture room. (See Tip 4, page 25.)

_____ _____   8. I am prepared to take notes at the start of the lecture. (See Tip 5, page 25.)

_____ _____   9. I ask questions during the lecture. (See Tip 6, page 25.)

_____ _____   10. I am able to focus my full attention on the lecture. (See Tip 7, page 25.)

_____ _____   11. I mark down on a piece of paper each time my mind begins to wander. (See Tip 8, page 25.)

Review the statements to which you have answered no. In addition, review the listening skills suggestions that you have already written to yourself on a separate sheet of paper. Then, do the following:

A. Prioritize the list of suggestions and select three behaviors that you can change that will help you to improve your listening during chemistry lecture.

B. Write these behavior changes in the space below.

C. Be specific about what you want to change, how you are going to make the change and when the new behavior will be implemented.

D. See if these changes help you better listen during a chemistry lecture.

1.

2.

3.

Do not try to make too many listening changes at one time because this can be overwhelming and lead to frustration. Once you have implemented the above three changes and seen improvement in your skills, we suggest that you again review areas in your listening that need improvement. Then go through steps A through D and watch for further improvement.

Tips for Note-Taking During Your Chemistry Lecture

As you read the following tips, write down on a sheet of paper anything you want to do differently to improve your note-taking.

1. Use *lots of 8 1/2" × 11" lined note paper* and keep it in a loose leaf binder. This allows you to

   a. Keep your notes organized and reduce the chance of their being lost (a potential disaster).

   b. Add extra sheets of paper to a section. This is especially useful for students when they do outside readings and find additional materials which help to clarify the lecture notes.

   c. Place the pass-out sheets next to the appropriate material in your notes instead of leaving them stuffed in your notebook or scattered around your study area, room, or campus.

2. *Use only one side of each piece of paper.* This provides free space to add to your notes as needed.

   a. There will be room to write in ideas or details taken from your textbook or other sources.

   b. You can use the added space to prepare short outlines or study guides that will help you study your notes.

3. *Leave a large margin* on the left (or right) side of the page.

   a. You will then be able to add comments about the material at a later time.

   b. You can write down any questions that arise during lecture. You can look up the answers later, discuss them in your study group, or ask the professor.

4. *Leave a lot of blank space* on each page of notes.

   a. You will find that it is easier to correct mistakes if there is room for corrections.

   b. Your notes will be easier to read because they will be more organized and not crammed together.

5. *Label your notes* (for example, "Chem 111"), date them, and number each page. This will be useful for later reference, for future studying, or if you accidentally drop your notes.

6. *Write legibly*. When you prepare for a test, you want to spend your time studying, not deciphering notes.

7. *Use abbreviations and symbols* to save time, but make sure you know what they mean, that they are correct and that the same abbreviations and symbols are used each time.

For example, it is much faster to write N_2 than nitrogen or mm instead of millimeter. In addition, using abbreviations and symbols will help you to learn chemistry more quickly.

8. Try to write down your notes in the following *outline form*:

Main points are listed at the margin.

Secondary points and supporting details are indented.

More subordinate material is further indented.

Definitions, for instance, should always start at the margin. Definitions in chemistry often include symbols, abbreviations, chemical formulas and chemical and mathematical equations. However, the format is still the same.

For example, a typical chemistry definition might look like this:

ACID   any substance that furnishes protons (H^+) or hydronium ions (H_3O^+) when added to water (H_2O).

$$HA + H_2O \rightarrow H_3O^+ + A^-$$

HCl, HBr, and HI are some common acids.

When a list of terms is presented, the heading should start at the margin, and each item in the series should be set in slightly from the margin.

Here is another organizational aid: When your professor moves from one idea or aspect of a topic to another, show this with blank spaces—skip a line or two or draw a line between topics.

Because of the nature of the material in chemistry, it is not always possible to use the outline techniques of indentation and blank space. However, they are the first steps toward organizing class material and should be used whenever possible.

9. *Watch for signals of importance.*

    a. *Write down whatever your professor puts on the board.* Ideally, print the material. If you don't have time to print, write legibly and put the letters "OB" in the margin to indicate that the material was written on the board. Later, when you review your notes, you will know which ideas your professor decided to emphasize. Material he or she took the time to put on the board is more likely to appear on the exams.

    b. If your professor emphasizes a point, write it down and note the emphasis in the margin, with a symbol you use consistently, such as "*" or "IMP." Examples of emphasis are when your professor says, "This is an important reason. . ."; or "The idea here is. . ."; or "Don't forget that. . ."; or "Pay special attention to. . ."; and so on.

c. If your professor repeats a point, assume it is important. In order to note it, you might write "R" in the margin for repeat. In chemistry, one way to repeat a point is to work several problems illustrating a concept. Write them all down.

d. If your professor's voice slows down, becomes louder, or otherwise signals that you should write down his or her exact words, needless to say, do so.

10. *Write down your professor's examples and mark them "EX."* The examples help you understand the abstract points. If you don't write examples down, you are likely to forget them later when they are needed to help make sense of an idea.

    a. In chemistry, many examples in the form of solved problems are used to explain a concept.

    b. A chemistry professor often uses examples similar to problems that appear on exams. These examples may also bring out new wrinkles to a concept not covered in the text.

11. When writing down *your own thoughts and questions* about a lecture, enclose them in parentheses, ( ) or brackets, [ ] or mark them with a "Q" or "?." This will allow you to separate your thoughts from those of the professor and enable you to easily identify questions you have about the lecture.

12. *Leave blank spaces, marked with an "L" for lost, for items or ideas you miss*.

    a. Right after class, ask another student or your professor to help you fill in the gaps.

    b. If you fall behind in note-taking, concentrate on getting down the material written on the board rather than the explanations. You may be able to get the explanations from another student, from the text, from your professor, or you may be able to figure it out by yourself.

13. *Do not try to write down every word the professor is saying*. Just write the key phrases and points. Complete explanations can be added later.

14. *Do not stop taking notes toward the end of a class*. Because of time spent answering questions, professors may have to hurry during the last minutes of class to finish a problem or topic. If this happens, be

ready to write as rapidly as possible. *Do not* get up and leave the lecture if the professor runs into overtime.

Also, be prepared to resist the fatigue that may sometimes settle in during a class. Remember that as a lecture proceeds, the possibility of losing attention increases.

You do not want to snap out of a daydream only to realize that your professor is halfway into an important idea that you haven't begun to write down.

15. *Edit your notes soon after class.* Within an hour after class, you may not be able to recall even half of what you heard during the lecture. Therefore, you must make your notes as clear as possible while they are still fresh in your mind. Going over your notes is a good review session and allows you to

   a. fill in the gaps.

   b. correct chemical symbols, formulas, etc.

   c. add symbols, comments, or other material that helps make your notes clear.

   d. get your notes ready for review before an exam.

   e. prepare for the next lecture.

16. *Remember, notes should help you study for tests.*

   a. Highlight (underline) your notes as you study.

   b. Write cue words or questions in the margin of your notes.

EXAMPLE OF POOR CHEMISTRY NOTES

Percent Composition — Call Sally about a date

Example:

The [molecular formula] for water is H_2O?
Determine the [molecular weight] — 1 mole water

$$H \times 2 \times 1.01 = 2.02$$
$$O \times 1 \times 16.00 = 16.02$$

[molar mass] =

[value] — 0.122

EXAMPLE OF GOOD CHEMISTRY NOTES

CH 111

9-27
Page 3

Per Cent Composition — tells how much
(%) of an element
is contained in
a compound

— significant
because ?

Example:

The molecular formula for water is
H_2O. Determine the mass % of
hydrogen in water.

1. Determine M.W. of H_2O

$2H = 2 \times 1.01 = 2.02g$

$O = 1 \times 16.00 = 16.00g$

$\overline{18.02g}$

Note-Taking Skills Check List

To help you clarify what note-taking changes you need to make, answer yes or no to the following statements. If necessary, review the appropriate tips for the "no" statements.

YES   NO

1. I keep my notes in a looseleaf binder. (See Tip 1, page 28.)

2. I use only one side of each piece of paper. (See Tip, page 28.)

3. I leave a large margin on each piece of paper. (See Tip 3, page 28.)

4. I leave a lot of blank space on each page. (See Tip 4, page 28.)

5. I label, date, and number each page. (See Tip 5, page 28.)

6. My notes are legible. (See Tip 6, page 29.)

7. I correctly use the formulas, symbols, abbreviations, and equations associated with chemistry. (See Tip 7, page 29.)

8. I write my notes in outline form. (See Tip 8, page 29.)

9. I write down all definitions. (See Tip 8, page 29.)

10. I use abbreviations and symbols to show that something was important (IMP or *), or written on the board (OB), or was an example (EX), or was repeated by the professor (R). (See Tip 9, page 30.)

11. I write down the professor's examples. (See Tip 10, page 31.)

12. I enclose my own thoughts and questions with ( ) or [ ], or indicate when I have a question with ? or Q by the questions. (See Tip 11, page 31.)

13. I leave blank spaces marked by an L when I get lost during a lecture. (See Tip 12, page 31.)

14. I write down everything the professor writes on the board. (See Tip 13, page 31.)

15. I take notes up until the end of lecture. (See Tip 14, page 31.)

16. I edit my notes as soon after each class as possible. (See Tip 15, page 32.)

17. I highlight my notes and write cue words or questions in the margins when studying for a test. (See Tip 16, page 32.)

_____   _____   **18.** I review my notes and recite the material using cue words and questions from the margin when preparing for a test. (See Tip 16, page 32.)

Review the statements to which you have answered no. In addition, review the suggestions for note-taking that you have already written to yourself on a separate sheet of paper. Then do the following:

A. Prioritize the list of suggestions and select three behaviors that you can change that will help you to improve your note-taking during chemistry lecture.

B. Write these behavior changes in the space below.

C. Be specific about what you want to change, how you are going to make the change, and when the new behavior will be implemented.

D. See if these changes help you take better notes during a chemistry lecture.

  1.

  2.

  3.

Do not try to make too many note-taking changes at one time because this can be overwhelming and lead to frustration. Once you have implemented the above three changes and have seen improvement in your skills, we suggest that you again review areas in your note-taking that need improvement. Then go through steps A through D and watch for further improvement.

Remember, attending lecture and taking good notes are critical to your success in chemistry. The professor sets the stage during the lectures by telling you what is important. Your notes will serve as a reminder for what to study. If you follow these simple guidelines, you may find that you enjoy attending lectures. You will understand the material during lecture and as a result your notes will make more sense when you review them.

4 READING YOUR CHEMISTRY TEXTBOOK

Reading a chemistry textbook can be challenging. When you first leafed through your textbook, you may have discovered many terms that you were unfamiliar with (for example, stoichiometry, Boyle's law, electronegativity, dipolar molecules). This new terminology can be overwhelming to the beginning chemistry student. But keep in mind that learning chemistry is like learning a foreign language. You must learn the new terms and formulas so that you can speak the language of chemistry.

If you never learn the meanings of stoichiometry, Boyle's law, electronegativity, or dipolar molecules, reading the chemistry textbook with understanding will be impossible.

Another difficulty in reading a chemistry text is that each chapter contains numerous problems that you must work and understand as you read. This is time-consuming and requires that you be *actively* involved in your reading. The homework problems at the end of each chapter are also necessary for you to practice what you've learned and to help you determine how well you understand the concepts presented.

There are several skills that you can develop to meet the challenge of reading a chemistry textbook. These skills will help you to

1. become a more efficient reader.

2. gain a better understanding of what you read.

3. retain more information.

In this chapter, we'll be giving you some helpful tips on building and improving chemistry textbook reading skills, but let's first take a look at some problems that chemistry students often experience while reading chemistry textbooks.

Problems Students Frequently Experience when Reading Their Chemistry Text

Read the following list of problem statements and place a check mark in the appropriate space to the left to indicate whether the statements are true for you.

YES NO

_____ _____ 1. I often don't remember what I've read.

_____ _____ 2. I often don't know the basic terms used in the text.

_____   _____   3. The problem examples in the text usually look easy, but when I try to do the ones at the end of the chapter, they are too difficult.

_____   _____   4. The text is so boring that I can't get into it.

_____   _____   5. It seems like there is too much to know in order to understand a chemistry text, and I can't do it.

_____   _____   6. I study a lot, but I seem to forget the important terms and formulas.

_____   _____   7. I often fall asleep when I'm reading my chemistry text or my mind begins to wander.

_____   _____   8. I seem to understand things fairly well right after I read a chapter, but when I go back to review before a test it seems like I've forgotten everything.

If you marked any of the above statements yes, you could benefit from improving your chemistry text reading skills. Read the following tips for improving your chemistry textbook reading skills.

If none of the above statements were true for you and you don't experience any problems with reading your chemistry textbook, skip ahead to the next chapter, "Chemistry Problem Solving."

Tips for Improving Your Chemistry Textbook Reading Skills

As you read these tips, you'll find it helpful to look at your chemistry textbook at the same time. Try going through a chapter using our recommended method.

As you read the following tips, write down on a sheet of paper anything you want to do differently to improve your reading of the chemistry textbook.

BEFORE YOU BEGIN READING YOUR FIRST TEXTBOOK ASSIGNMENT

1. *Skim through your textbook, look at the table of contents, and read the preface.* Think about how helpful it is to look at a map of a new city before driving in it. The same is true with reading. Skimming a chapter will give you a framework to understand the material once you begin reading. The author may use the preface or sections of the first chapter to give you important information about what will be included in

the text, tips on reading and studying chemistry, and sometimes advice on working problems.

2. *Look at the appendices.* While skimming your text, pay careful attention to the appendices. These often contain tables of constants and reference data, mathematical reviews, and answers to problems. This is important information that you'll use when working problems and it's helpful to know where you can find it before you begin.

3. *Skim each chapter before you begin reading it.*

   a. Begin with an overview of the chapter by reading and thinking about the subtitles which are usually in **bold** print.

   b. After you've read and thought about the first subtitle (asking yourself questions such as, How does this relate to the chapter title? What is this about?), read the first sentence of each paragraph under that subtitle and once again, think about it, ask yourself questions, make a mental note of any unfamiliar terms, then go on to the second subtitle and go through the same process.

   c. Reading the first sentence of each paragraph under a subtitle won't tell you everything about that section, but it will give you an idea about the contents of that section, and what's going to be important in the chapter. In this phase of your reading, you are not expected to understand everything you read. Many of the terms and mathematical manipulations are likely to be unfamiliar to you, but by glancing at them and giving them some thought, your later work of trying to understand them will be easier.

4. *Examine tables, figures, and diagrams under each subtitle while skimming the chapter.*

   a. Chemistry textbooks contain many tables and figures that are extremely helpful in illustrating points. An author sometimes takes a page or two to explain verbally what may be illustrated in one small diagram. For example, it may take an author several paragraphs to explain molecular shapes, but it is the following diagram that clearly depicts a geometry.

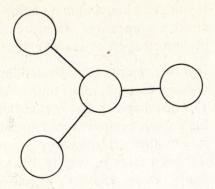

b. By examining the tables, figures, and diagrams before reading a chapter, you will become acquainted with the chapter highlights and will be able to go through the verbal explanations faster and with a better understanding.

c. People tend to remember what they see first. When you look at an illustration in advance, your first exposure is to the important information, without the verbal details interfering.

5. *When skimming the chapter, pay special attention to words that are underlined or written in boldface or italics.* The author thought these words were important and that's why they are written to stand out.

YOU'RE NOW READY TO BEGIN READING THE CHAPTER

6. *Read the text sentence by sentence.* If you quickly read over an entire paragraph or chapter you're likely to have difficulties understanding the material. Like a foreign language, each sentence in your chemistry text takes some work at understanding.

a. Stop periodically and ask yourself what you've learned thus far. Don't continue until you feel confident about your understanding of each concept.

b. If you encounter a term that you are unsure of, find out what it means before you continue. For example, you may be unclear about the term *anion* that keeps recurring in the chapter you are reading. It will be very difficult for you to understand the material if you do not understand this term. Read the following:

"Anions and cations are oppositely charged."

If you don't know what an anion (or cation) is, chances are that the above sentence was incomprehensible to you. You will not retain the information unless you understand what you read.

    c. If your attempts at understanding a new term, concept, or problem are unsuccessful, make note of that (make up a 3 × 5 card, see Tip 11, this chapter) and consult another introductory chemistry text for a fresh perspective. You can also seek help from your professor, another student, study group, or tutor. But remember that *you* will learn more if, in spite of your frustration, you work at trying to figure out the solution yourself.

    d. By continually asking yourself questions, testing yourself, and working to figure things out, you're *Actively Reading* the material. this will facilitate your learning and retention.

7. *Work through the example problems.* Look at the following solved problem. First, simply read through the problem and solution. Then, cover up the solution and try to solve the problem.

PROBLEM   The density of carbon tetrachloride is 1.595 g/cm^3. How many milliliters are contained in 28.00 grams?

SOLUTION

$$(28.00 \text{ g CCl}_4) \left(\frac{\text{cm}^3}{1.595 \text{ g}} \right) = 17.55 \text{ cm}^3$$

Remember, a cubic centimeter (cm^3) is the same as a milliliter.

You can see that it is deceiving to *read* and not *work* through problems. The ease with which you can read problems makes it seem like you understand the material and can easily solve the problems, when in reality, you cannot. This can make doing homework problems even more frustrating. Learning from a chemistry text definitely means that you have to work while you read.

8. *Use the illustrations in the text.* Forming a visual image of a concept, definition, procedure, or an apparatus, especially in science, is extremely helpful in being able to understand that concept and to remember it. For example, an author could explain in great detail what a buret looks like and it still may not be clear to you. However, if you look at a diagram of a buret, you will have no doubt in your mind as to what it looks like.

BURET

If a concept is presented where there is no illustration in your text, try to *visually create* your own illustration in your mind. Scientists often use this type of visual imagery because it enhances the understanding of scientific concepts. This accounts for the numerous tables and figures in a chemistry textbook. *Use them or make up your own.*

9. *Underline your textbook.* Underlining helps you remember important material. You should first read a paragraph and then go back and underline important points and definitions. *Do not underline* until you've read the paragraph or you'll underline items that are unimportant.

   a. You may want to write key words in the margins to use later on as cues in review. Even writing personal comments such as "Boring!" may help you to be more involved with your reading.

   b. Remember, having a book without marks in it will not make you a star student.

10. *Outline the chapters.* One of the best ways to study chemistry is to outline the text before you attend lecture, and following lecture, to outline your lecture notes. You can then add significant material from the chapter to your lecture outlines. Be careful not to make mistakes when you're copying material from one place to another.

11. *Use 3 × 5 cards to write down anything that needs to be memorized or that you need to give extra attention to.* As you read the chapter, write down unfamiliar terms, chemical and mathematical formulas, and equations on one side of the card, with definitions, and formulas, or equations on

the other side. Be sure to write down only one piece of information per card.

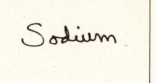

EXAMPLE OF 3 × 5 CARD

a. You will be able to learn the terminology and formulas by carrying the cards with you and glancing at them whenever you get a chance.

b. Remember that in chemistry it is very important that you know the terms and formulas exactly.

12. *Recite aloud, in your own words, the important concepts, definitions, and formulas from a paragraph you have just read.*

a. Recitation is most effective if you use the following technique: Cover up a paragraph, use your key words if written in the margin or on your 3 × 5 cards as cues, and repeat the main points of the paragraph, the definitions, formulas, and equations. For example, try to explain to yourself what a mole is, chemically speaking. If you are unable to recite with understanding, reread the chapter section that explains what a mole is, rework the examples, and then try reciting again. Follow this procedure whenever you have difficulties with recitation.

b. *Do not* substitute rereading a paragraph for recitation. *It will not work.* Recitation is much more effective for keeping information in your memory. In addition, it promotes concentration, helps you to understand the next paragraph, ensures that terminology and formulas are learned accurately, and provides you with immediate feedback on how you're doing. Not only does it help you know the material, it helps you *know* that you know the material. This boosts your confidence and sense of mastery for tests and for your general satisfaction.

c. Research has consistently shown that the greater the proportion of reciting time, the greater the learning and retention. This is especially true in chemistry, where accuracy in learning terms and formulas is crucial.

13. *Make notes on your recitation*. These notes can be very helpful to you as you review. Be sure to record in these notes any difficulties you have.

AFTER YOU'VE FINISHED READING THE CHAPTER YOU'RE READY FOR THE FOLLOWING

14. *Review the chapter*.

a. *Immediate Review*. As soon as you've finished reading the chapter sentence by sentence, begin your review. Cover the first page of the chapter with a piece of paper, leaving any marginal notes exposed. Use marginal or recitation notes, or your textbook outline, as cues to recite aloud in your own words the ideas, concepts, terms, and formulas *from that page*. Then remove the paper and check yourself. Correct any mistakes or omissions. Go through each page of the chapter this way. It will help you remember what you learned and build your confidence as you master the material.

b. *Later Review*. Immediate review as soon as you've finished reading a chapter is important but you also need to continue the review process thoroughly and often. Use the immediate review process of covering each page, using cue words to recite the material in your own words, checking yourself, and making corrections. If there is any subsection that you cannot review in this way, don't continue. Instead go back and try again to read, work the problems, recite, and make notes on your recitation. When you can go through the subsection using these four steps with mastery, you're ready to continue on to the next page.

c. This review process is excellent for learning and remembering difficult material such as chemistry. It is an active learning process which allows you to spend most of your time on the material you don't know. Passively rereading a chapter or going over your notes results in skipping over material you don't understand, focusing mainly on what comes most easily for you. This review procedure forces you to apportion your study time to the different sections of a chapter according to how well you understand them. You then spend the most time on the material that is most difficult for you.

15. *Work the assigned problems.* It cannot be emphasized enough that one of the best ways to learn chemistry is to do problems, and lots of them. Problem solving in chemistry is important enough to warrant an entire chapter in this book. If you are having difficulties with problem solving, see the next chapter, "Chemistry Problem Solving."

Chemistry Textbook Reading Skills Check List

To help you clarify what changes you need to make to improve your chemistry textbook reading skills, answer yes or no to the following statements. If necessary, review the appropriate tips for "no" statements.

YES NO

_____ _____ **1.** Before reading my first chemistry assignment. I skim the text. (See Tip 1, page 39.)

_____ _____ **2.** I am familiar with my chemistry textbook appendices. (See Tip 2, page 40.)

_____ _____ **3.** Before I read a chemistry text chapter I skim it. (See Tip 3, page 40.)

_____ _____ **4.** While skimming a chapter, I become familiar with the tables, figures, and diagrams. (See Tip 4, page 40.)

_____ _____ **5.** When skimming a chapter, I carefully read any words that are underlined or written in boldface or italics. (See Tip 5, page 41.)

_____ _____ **6.** When reading a chapter, I carefully read each sentence. (See Tip 6, page 41.)

_____ _____ **7.** When I encounter a term that I am unsure of, instead of skipping over it, I find out what it means. (See Tip 6, page 41.)

_____ _____ **8.** I am actively involved with my reading. (See Tip 6, page 41.)

_____ _____ **9.** I work through the example problems. (See Tip 7, page 42.)

_____ _____ **10.** I use the illustrations in the text and I create my own. (See Tip 8, page 42.)

_____ _____ **11.** I underline in my chemistry textbook. (See Tip 9, page 43.)

_____ _____ **12.** I outline my textbook chapters and add significant material from those outlines to my lecture outlines. (See Tip 10, page 43.)

_____ _____ **13.** I use 3 × 5 cards to facilitate my memorization work. (See Tip 11, page 43.)

_____ _____ **14.** I use recitation after reading each paragraph. (See Tip 12, page 44.)

_____ _____ **15.** I make notes on my recitation. (See Tip 13, page 45.)

_____ _____ **16.** I review a chapter immediately after I have finished reading it. (See Tip 14, page 45.)

_____ _____ **17.** I use recitation when I review. (See Tip 14, page 45.)

_____ _____ **18.** I frequently review chapters I have read. (See Tip 14, page 45.)

_____ _____ **19.** I work all of the assigned problems. (See Tip 15, page 46.)

Review the statements to which you answered no. In addition, review the suggestions for chemistry reading skills improvement you have already written on a separate sheet of paper. Then, do the following:

A. Prioritize these lists and select three behaviors you can change that will help you to improve your reading of the chemistry text.

B. Write these behavior changes in the space below.

C. Be specific about what you want to change, how you are going to make the change, and when the new behaviors will be implemented.

D. See if these changes help you better read the chemistry text.

   1.

   2.

   3.

Do not try to make too many changes at one time because this can be overwhelming and lead to frustration. Once you have implemented the above three changes and have seen improvement in your reading skills, we suggest that you again review your chemistry reading habits. Then go through steps A through D and watch for further improvement.

5 CHEMISTRY PROBLEM SOLVING

> **Is working chemistry problems as difficult as it seems?**
> **?**

Chemistry is taught through problem solving and you cannot learn it any other way. One of the biggest stumbling blocks in solving chemistry problems is the analysis of the problems themselves. Analysis involves picking a problem apart to understand what is being asked and then working to obtain a solution. There are several things you must keep in mind when solving chemistry problems in order to make this process less painful. In this chapter, we will be examining just what those things are.

Let's first take a look at some problems chemistry students frequently report with chemistry problem solving.

Problem-Solving Difficulties

Read the following list of problem statements and place a check mark in the appropriate space to the left to indicate whether the statements are true for you.

YES   NO

1. I have trouble reading and understanding chemistry problems.

2. I never know where to begin when solving the problems.

3. I feel stupid when I work on my chemistry homework.

_____ _____ **4.** Working on chemistry problems makes me hate chemistry.

_____ _____ **5.** I get frustrated when I'm doing my chemistry homework.

_____ _____ **6.** I can't make sense of my solution sheets.

_____ _____ **7.** I feel lost without the solutions manual.

_____ _____ **8.** I can't understand significant figures.

_____ _____ **9.** I have no faith in my ability to solve chemistry problems.

_____ _____ **10.** I never know if I have solved the problem correctly until I check the solutions manual.

If you marked any of the above statements yes, you could benefit from improving your problem-solving skills. Read the following tips for chemistry problem solving.

If none of the statements were true for you and you don't experience any difficulties with problem solving, skip ahead to the next chapter, "The Chemistry Laboratory."

Tips for Analyzing Chemistry Problems

As you read the following tips, write down on a sheet of paper anything you want to do differently to improve your problem-solving skills.

Read the following problem and refer to it as you read through the tips section.

EXAMPLE
PROBLEM 1

Determine the number of grams and moles of sodium carbonate that can be prepared from 88.0 grams of sodium chloride and excess calcium carbonate according to the following equation:

$$2\,NaCl(s) + CaCO_3(s) \rightarrow Na_2CO_3(s) + CaCl_2(s)$$

1. Read the entire problem above and make note of everything you are unfamiliar with.

   a. You must read the entire problem before you try to solve it. Too often students start to work when they encounter the first numbers while reading the problem, before they have an idea as to what is being asked or how they are going to solve it.

   b. Underline anything you are unfamiliar with.

   i. Do you know the formulas for sodium chloride ($NaCl$) and sodium carbonate (Na_2CO_3)? If not, you will need to review the section on writing chemical formulas.

   ii. Do you know what a mole is? If the answer is no, look up the definition of a mole, make up a 3×5 card and go over this definition until it becomes second nature to you. Remember that you do not want to memorize this definition, you want to understand it and be able to apply it. The concept of a mole is extremely important in chemistry and you will find that it recurs often throughout the entire course. Keep in mind that chemistry builds on knowledge acquired as the course proceeds.

2. *Determine the problem topic.* For example, ask yourself

   a. Is this a gas law problem? The answer is no.

   b. Does this problem have anything to do with atomic structure? The answer is no.

   c. Is this a stoichiometry problem? The answer is yes.

If you are doing a problem at the end of the chapter, you will have some idea as to the principle involved in the problem because of the subject matter covered in the chapter. On an exam it may not be so easy. Several chapters may be covered or several principles may be incorporated into one question (for example, there may be a gas law problem involving stoichiometry).

3. *Write down all the information given.* In the example problem this included the balanced equation, 88.0 grams of sodium chloride and excess calcium carbonate. One way to write down this information would be

KNOW:

$$2\,NaCl\,(s) \;+\; CaCO_3\,(s) \longrightarrow Na_2CO_3\,(s) \;+\; CaCl_2\,(s)$$

$$88.0\,g \qquad\qquad EXCESS$$

4. *Determine what you are trying to solve for.* In this particular problem, you will be calculating the number of grams and moles of sodium carbonate that can be prepared from 88.0 grams of sodium chloride and excess calcium carbonate. Your problem sheet may now look like this:

KNOW:

$$2\,NaCl\,(s) \;+\; CaCO_3\,(s) \longrightarrow Na_2CO_3\,(s) \;+\; CaCl_2\,(s)$$

$$88.0\,g \qquad EXCESS \qquad\qquad ?$$

WANT:

GRAMS AND MOLES OF SODIUM CARBONATE

You may find that what you are looking for in a problem is not always so obvious. You're likely to make some false starts, but with practice you will become better at problem solving. Don't get discouraged. Look at this as a learning experience and you will be able to monitor your growth and improvement in solving chemistry problems.

5. Plan your method of attack. Before you jump in and try to manipulate numbers and race along to get to the answer, sit back and think about how you are going to solve the problem.

   a. Think about the chemical concepts illustrated in a problem. In this particular problem, you are learning to use chemical formulas and equations, along with the concept of a mole, in order to do a stoichiometric calculation.

   b. Think about the mathematical formulas and manipulations to be used in solving the problem. For a stoichiometric calculation you could set something up like this:

CALCULATIONS

GRAMS OF ⟶ MOLES OF ⟶ MOLES OF ⟶ GRAMS OF
Na Cl   Na Cl   Na₂ CO₃   Na₂ CO₃

┌─────────────────────────────────────┐
│ OR AN ALTERNATE SOLUTION │
└─────────────────────────────────────┘

(GRAMS OF NaCl) (M.W. OF NaCl) = MOLES OF NaCl

(MOLES OF NaCl) (BALANCED EQUATION) = MOLES OF Na₂ CO₃

(MOLES OF Na₂ CO₃) (M.W. OF Na₂ CO₃) = GRAMS OF Na₂ CO₃

THE UNITS ON MOLECULAR WEIGHT ARE g/mol

6.   *Determine if there is enough information given* for you to work the problem. For example, in the above problem you are given the number of grams of sodium chloride but you will need to look up or determine the molecular weight of sodium chloride and sodium carbonate. Quite often you may have to look up constants in the appendices or in the tables in the chapters. You could add the following to your worksheet:

NEED:

MOLECULAR WEIGHT OF NaCl = 58.44 g/mol

MOLECULAR WEIGHT OF Na₂ CO₃ = 105.99 g/mol

7. *Determine if there is too much information.* Although this doesn't apply to this example, we will later give you an example where irrelevant information is stated in the problem.

8. *Solve the problem in an organized and neat fashion* and include all units in your solution. If you make a mistake it will be easier to determine where you went wrong. If you solve it correctly, it will be easier for you to review the material at a later date. Keep in mind that when you are taking an exam your solution must be neat and organized so that the professor can follow it or you will get very little, if any, credit. Your problem sheet may now look like this:

KNOW:

$$2\,NaCl_{(s)} + Ca\,CO_{3\,(s)} \longrightarrow Na_2\,CO_{3\,(s)} + Ca\,Cl_{2\,(s)}$$

    88.0 g   EXCESS

WANT:

GRAMS AND MOLES OF SODIUM CARBONATE

CALCULATIONS:

GRAMS OF $\longrightarrow$ MOLES OF $\longrightarrow$ MOLES OF $\longrightarrow$ GRAMS OF
   NaCl   NaCl $Na_2\,CO_3$ $Na_2\,CO_3$

OR AN ALTERNATE SOLUTION

(GRAMS OF NaCl)(M.W. OF NaCl) = MOLES OF NaCl

(MOLES OF NaCl)(BALANCED EQUATION) = MOLES OF $Na_2\,CO_3$

(MOLES OF $Na_2\,CO_3$)(M.W. OF $Na_2\,CO_3$) = GRAMS OF $Na_2\,CO_3$

THE UNITS ON MOLECULAR WEIGHT ARE g/mol

NEED:

MOLECULAR WEIGHT OF NaCl = 58.44 g/mol

MOLECULAR WEIGHT OF $Na_2\,CO_3$ = 105.99 g/mol

SOLUTION:

$$(88.0g \; NaCl)\left(\frac{1 \; mol \; NaCl}{58.44 \; g \; NaCl}\right)\left(\frac{1 \; mol \; Na_2CO_3}{2 \; mol \; NaCl}\right)\left(\frac{105.99 \; g \; Na_2CO_3}{1 \; mol \; Na_2CO_3}\right)$$

$$= 79.8 \; g \; Na_2CO_3$$
*

OR AN ALTERNATE SOLUTION

$$(88.0g \; NaCl)\left(\frac{1 \; mol \; NaCl}{58.44 \; g \; NaCl}\right) = 1.51 \; mol \; NaCl$$

$$(1.51 \; mol \; NaCl)\left(\frac{1 \; mol \; Na_2CO_3}{2 \; mol \; NaCl}\right) = 0.755 \; mol \; Na_2CO_3$$

$$(0.755 \; mol \; Na_2CO_3)\left(\frac{105.99 \; g \; Na_2CO_3}{1 \; mol \; Na_2CO_3}\right) = 80.0 \; g \; Na_2CO_3$$

*The student did not completely solve the problem. He or she forgot to solve for the number of moles of Na_2CO_3. Don't make mistakes like this on an exam.

Note: There are two different ways to solve the problem. Each gives the correct answer.

9. *If you are having difficulty* solving the problem, do the following:

   a. Reread the section in the text that covers the topic.

   b. Review the example problems in the text.

   c. Review your class notes.

   d. Consult supplementary texts.

   e. Check your mathematics.

10. If you cannot solve the problem after 15–20 minutes, you have missed something important. Do the following:

   a. Look at the solutions manual but make a note to yourself, as you'll need to spend more time on this type of problem.

   b. Discuss the problem with your study group.

   c. Seek help from your professor or a tutor.

   It is important that you understand why you had difficulties. Don't let this slide by, because this same type of problem will keep recurring at later times.

11. *Include the units* in your solution. This is important because it will help you better understand chemistry and make sense of your answers. You will also find that you will be expected to use units on exams, and if you don't include them when you are solving the homework problems, it will be very difficult for you to start using them during an exam. Look at the following examples:

CALCULATION OF THE NUMBER OF MOLES OF NaCl

CORRECT CALCULATION:

$$(88.0 \text{ g NaCl})\left(\frac{1 \text{ mol NaCl}}{58.44 \text{ g NaCl}}\right) = 1.51 \text{ mol NaCl}$$

INCORRECT CALCULATION:

$$(88.0 \text{ g NaCl})\left(\frac{58.44 \text{ g NaCl}}{1 \text{ mol NaCl}}\right) = 5140 \text{ g}^2 \text{ mol}^{-1} \text{ NaCl}$$

You could tell just from the units alone that the second calculation is incorrect. Remember, you want moles of NaCl and not $g^2 \text{ mol}^{-1}$ of NaCl.

12. *Check your answer* to see if it makes sense. Look at the sample calculation in Tip 11. If you determine that there are 5140 moles of NaCl in 88.0 grams of NaCl, you are incorrect. You will discover that after you do a few of the problems with understanding you will develop some insight into what the answer should be and it will become easier for you to detect errors. *Persevere.*

13. *Watch your significant figures*. Review the section on significant figures in your text and pay attention to their use.

<u>INCORRECT USE OF SIGNIFICANT FIGURES:</u>

$$(88.0 \text{ g NaCl})\left(\frac{1 \text{ mol NaCl}}{58.44 \text{ g NaCl}}\right) = 1.505817933 \text{ mol NaCl (from calculator)}$$

By using significant figures correctly, the answer is 1.50 moles of NaCl. Remember, the rule for multiplication and division is that your answer is expressed in the same number of significant figures as the least number of significant figures in any one value in your calculation.

You will be expected to use significant figures correctly on exams. Misuse of them may result in the loss of points which can make a difference of a letter grade in your final grade.

14. *Pay attention to little mistakes*. If you find that you repeatedly make the same mistakes (for example, multiplying by the molecular weight instead of dividing when you are solving for the number of moles) when you are doing your homework, you will also make the same mistakes on the exams. Make note of your little mistakes either on a separate piece of paper or next to the solved problem and pay attention to this source of error. Little mistakes will cost you dearly on exams.

Our example problem 1 is straightforward. There is no mystery as to what is being asked and most students will have little difficulty in solving it. Let's complicate the problem a little by stating it in a form that is similar to that seen for chemistry homework problems and on exams.

Read the following problem and work through the tips. Keep in mind that this is a different problem from before, and there will be new problem-solving techniques.

EXAMPLE PROBLEM 2

Sodium chloride, common table salt, is of much concern today as a possible health hazard. Sodium chloride is also important in the chemical industry and one of its uses is in the preparation of calcium chloride according to the following equation:

$$2 \text{ NaCl(s)} + \text{CaCO}_3\text{(s)} \rightarrow \text{Na}_2\text{CO}_3\text{(s)} + \text{CaCl}_2\text{(s)}$$

Calcium chloride is used in the winter on the highways to melt ice. How many moles and grams of sodium carbonate can be prepared from 88.0 grams of sodium chloride and 88.0 grams of calcium carbonate? The melting point of calcium carbonate is 1339°C.

REVIEW OF TIPS FOR ANALYZING CHEMISTRY PROBLEMS FOR EXAMPLE PROBLEM 2

1. *Read* the entire problem. Make note of and underline everything you are unfamiliar with.

   Do you know the formulas for sodium chloride (NaCl), calcium carbonate ($CaCO_3$), and sodium carbonate (Na_2CO_3)? If not, you will need to review the section on writing chemical formulas.

   You may need to review the definition of a mole.

2. *Determine the problem topic.* For example, ask yourself:

   a. Is this a gas law problem? The answer is no.

   b. Does this problem have anything to do with atomic structure? The answer is no.

   c. Is this a stoichiometry problem? The answer is yes. This is still a stoichiometry problem but with a limiting reagent complication.

3. *Write down all the information given.* In this problem you are given the balanced equation, 88.0 grams of both sodium chloride and calcium carbonate and the melting point of calcium carbonate. You are told that sodium chloride can cause health problems, but this information doesn't help you to obtain the solution. In addition, it is not important to the problem that sodium chloride is used in industry or that calcium chloride is used on the highways in the winter. Quite often students become confused when they have to wade through this extraneous material. It is with experience that you will be able to filter out material which is not relevant to the problem. Your problem sheet might look like the following:

   KNOW:

   $$2\,NaCl_{(s)} + CaCO_{3(s)} \longrightarrow Na_2CO_{3(s)} + CaCl_{2(s)}$$

   $$\underline{88.0g} \qquad\quad 88.0g$$

   MELTING POINT OF $CaCO_3$ = 1339°C

4. *Determine what you are trying to solve for.* You are still trying to determine the number of grams and moles of sodium carbonate that can be
   prepared from the given amounts of the starting materials. The new
   wrinkle here is the limiting reagent concept.

KNOW:

$$2\,NaCl\,(s) \;+\; CaCO_3\,(s) \;\longrightarrow\; Na_2CO_3\,(s) \;+\; CaCl_2\,(s)$$

$$\underline{88.0\,g \qquad\qquad 88.0\,g}$$

$$\underline{MELTING\ POINT\ OF\ CaCO_3 = 1339°C}$$

WANT:

GRAMS AND MOLES OF SODIUM CARBONATE

5. *Plan your method of attack.*

   a. Think about the chemical concepts illustrated in a problem. In this
      particular problem, you are learning to use chemical formulas and
      equations, along with the concept of a mole, in order to do a Lim-
      iting reagent stoichiometric calculation.

   b. Think about the mathematical formulas and manipulations to be
      used in solving the problem. For a stoichiometric calculation with
      a limiting reagent complication you could set something up like
      this:

CALCULATIONS:

LIMITING REAGENT

| GRAMS OF $CaCO_3$ | $\longrightarrow$ | MOLES OF $CaCO_3$ | $\longrightarrow$ | MOLES OF Na_2CO_3 | $\longrightarrow$ | GRAMS OF Na_2CO_3 |
| GRAMS OF $NaCl$ | $\longrightarrow$ | MOLES OF $NaCl$ | $\longrightarrow$ | MOLES OF Na_2CO_3 | $\longrightarrow$ | GRAMS OF Na_2CO_3 |

OR AN ALTERNATE SOLUTION

(GRAMS OF NaCl) (M.W. OF NaCl) = MOLES OF NaCl

(MOLES OF NaCl) (BALANCED EQUATION) = MOLES OF Na_2CO_3

(MOLES OF Na_2CO_3) (M.W. OF Na_2CO_3) = GRAMS OF Na_2CO_3

AND

(GRAMS OF $CaCO_3$)(M.W. OF $CaCO_3$) = MOLES OF $CaCO_3$

(MOLES OF $CaCO_3$) (BALANCED EQUATION) = MOLES OF Na_2CO_3

(MOLES OF Na_2CO_3) (M.W. OF Na_2CO_3) = GRAMS OF Na_2CO_3

THE UNITS ON MOLECULAR WEIGHT ARE g/mol

6. *Determine if there is enough information* given for you to work the problem. For this problem you will have to determine the molecular weights of sodium chloride, calcium carbonate, and sodium carbonate. You could add the following to your problem sheet:

NEED:

MOLECULAR WEIGHT OF NaCl = 58.44 g/mol

MOLECULAR WEIGHT OF $CaCO_3$ = 100.09 g/mol

MOLECULAR WEIGHT OF Na_2CO_3 = 105.99 g/mol

7. *Determine if there is too much information.* In this problem, you are given the melting point of calcium carbonate which is not needed for the solution. You will find that "extra" information is *not* usually added to chemistry problems, especially at the introductory level. You will also discover that you may try to incorporate the given melting point into your calculations or be in a quandary as to what to do with it.

8. *Solve the problem in an organized and neat fashion* and include all units in your solution. Your problem sheet may now look like this:

KNOW:

$$2 NaCl(s) \; + \; CaCO_3(s) \longrightarrow Na_2CO_3(s) \; + \; CaCl_2(s)$$

$$\underline{88.0g} \qquad\qquad 88.0g$$

MELTING POINT OF $CaCO_3$ = $1339°C$

WANT:

GRAMS AND MOLES OF SODIUM CARBONATE

CALCULATIONS:

LIMITING REAGENT

GRAMS OF $CaCO_3$ $\longrightarrow$ MOLES OF $CaCO_3$ $\longrightarrow$ MOLES OF Na_2CO_3 $\longrightarrow$ GRAMS OF Na_2CO_3

GRAMS OF $NaCl$ $\longrightarrow$ MOLES OF $NaCl$ $\longrightarrow$ MOLES OF Na_2CO_3 $\longrightarrow$ GRAMS OF Na_2CO_3

OR AN ALTERNATE SOLUTION

(GRAMS OF NaCl) (M.W. OF NaCl) = MOLES OF NaCl

(MOLES OF NaCl) (BALANCED EQUATION) = MOLES OF Na_2CO_3

(MOLES OF Na_2CO_3) (M.W. OF Na_2CO_3) = GRAMS OF Na_2CO_3

AND

(GRAMS OF $CaCO_3$)(M.W. OF $CaCO_3$) = MOLES OF $CaCO_3$

(MOLES OF $CaCO_3$)(BALANCED EQUATION) = MOLES OF Na_2CO_3

(MOLES OF Na_2CO_3)(M.W. OF Na_2CO_3) = GRAMS OF Na_2CO_3

THE UNITS ON MOLECULAR WEIGHT ARE g/mol

<u>NEED</u>:

MOLECULAR WEIGHT OF NaCl = 58.44 g/mol

MOLECULAR WEIGHT OF $CaCO_3$ = 100.09 g/mol

MOLECULAR WEIGHT OF Na_2CO_3 = 105.99g/mol

SOLUTION:

* $(88.0 g \; NaCl)\left(\dfrac{1 mol \; NaCl}{58.44 g \; NaCl}\right)\left(\dfrac{1 mol \; Na_2CO_3}{2 mol \; NaCl}\right)\left(\dfrac{105.99 g \; Na_2CO_3}{1 mol \; Na_2CO_3}\right) = 79.8 g \; Na_2CO_3$

$(88.0 g \; CaCO_3)\left(\dfrac{1 mol \; CaCO_3}{100.09 g \; CaCO_3}\right)\left(\dfrac{1 mol \; Na_2CO_3}{1 mol \; CaCO_3}\right)\left(\dfrac{105.99 g \; Na_2CO_3}{1 mol \; Na_2CO_3}\right) = 93.2 g \; Na_2CO_3$

THE L.R. IS NaCl BECAUSE IT PRODUCES THE LEAST NUMBER OF GRAMS OF Na_2CO_3

AN ALTERNATE SOLUTION

$(88.0 g \; NaCl)\left(\dfrac{mol \; NaCl}{58.44 g \; NaCl}\right) = 1.51 mol \; NaCl$

$(1.51 mol \; NaCl)\left(\dfrac{1 mol \; Na_2CO_3}{2 mol \; NaCl}\right) = 0.755 mol \; Na_2CO_3$

$(0.755 mol \; Na_2CO_3)\left(\dfrac{105.99 g \; Na_2CO_3}{1 mol \; Na_2CO_3}\right) = 80.0 g \; Na_2CO_3$

AND

$(88.0 g \; CaCO_3)\left(\dfrac{1 mol \; CaCO_3}{100.09 g \; CaCO_3}\right) = 0.879 mol \; CaCO_3$

$(0.879 mol \; CaCO_3)\left(\dfrac{1 mol \; Na_2CO_3}{1 mol \; CaCO_3}\right) = 0.879 mol \; Na_2CO_3$

$(0.879 mol \; Na_2CO_3)\left(\dfrac{105.99 g \; Na_2CO_3}{1 mol \; Na_2CO_3}\right) = 93.1 g \; Na_2CO_3$

L.R. IS NaCl

* SEE PAGE 65 FOR EXPLANATION

9. If you are having difficulty solving the problem, do the following:

    a. Reread the section in the text that covers a particular topic.

    b. Review the example problems in the text.

    c. Review your class notes.

    d. Consult the supplementary texts.

    e. Check your mathematics.

If you cannot solve the problem after 15–20 minutes, you have missed something important. Do the following:

    a. Look at the solutions manual but make note of this problem.

    b. Discuss the problem with your study group.

    c. Seek help from your professor or a tutor.

It is important that you understand why you had difficulty with this problem.

10. *Include the units* in your solution. Look at the following examples:

<u>CALCULATION OF THE NUMBER OF GRAMS OF Na_2CO_3</u>

<u>CORRECT CALCULATION:</u>

$$(0.755 \text{ mol } Na_2CO_3) \, \frac{105.99 \text{ g } Na_2CO_3}{1 \text{ mol } Na_2CO_3} = 80.0 \text{ g } Na_2CO_3$$

<u>INCORRECT CALCULATION:</u>

$$(0.755 \text{ mol } Na_2CO_3) \left(\frac{1 \text{ mol } Na_2CO_3}{105.99 \text{ g } Na_2CO_3} \right) = \frac{7.12 \times 10^{-3} \, (\text{mol } Na_2CO_3)^2}{g}$$

Remember: You want grams of Na_2CO_3, not $(\text{mol}^2/g) \, Na_2CO_3$.

11. *Check your answer* to see if it makes sense. Look at the sample calculations in Tip 10. If you determined that there are 7.12×10^{-3} grams of Na_2CO_3 in 0.755 mol of Na_2CO_3, you are incorrect.

12. *Watch your significant figures.*

    Using significant figures correctly, the answer is 80.0 grams of Na_2CO_3.

    INCORRECT USE OF SIGNIFICANT FIGURES

    $$(0.755 \text{ mol } Na_2CO_3) \left(\frac{105.99 \text{ g } Na_2CO_3}{1 \text{ mol } Na_2CO_3} \right) = 80.02245 \text{ g (from calculator)}$$

13. Pay attention to little mistakes.

TWO ADDITIONAL TIPS WHICH WILL HELP YOU SOLVE PROBLEMS

14. *Make a list of important formulas* and how they are used. For example, it is important to know that the ideal gas law is $PV = nRT$. In addition, you must know what each symbol represents, the units on each symbol and the situations in which this formula can be used. You could make a table that looks like this:

Symbol	Meaning	Units
P	pressure	atmospheres
V	volume	liter
n	moles	moles
R	gas constant	$0.0821 \text{ L} \cdot \text{atm} (\text{K} \cdot \text{mol})^{-1}$
T	temperature	Kelvin

The ideal gas law is used to solve problems involving gases, not liquids and solids.

15. *Review any mathematical manipulations* that you have difficulty with. Look in the appendices of your text or a basic algebra book for the manipulations that are used in chemistry. If you have repeated difficulty with them you will want to write them on a separate sheet of paper for quick reference. For example, when you are doing an ideal gas law problem, you often are trying to isolate one variable. If you have difficulty with this, write it out.

EXAMPLE

$PV = nRT$ To solve for T, divide both sides of the equation by nR.

$\dfrac{PV}{nR} = \dfrac{nRT}{nR}$ The nR cancels from the right side of the equation.

$\dfrac{PV}{nR} = T$

PROBLEM-SOLVING SKILLS CHECK LIST

To help you clarify what problem-solving changes you need to make, answer yes or no to the following statements. If necessary, review the appropriate tips for the "no" statements.

YES   NO

_____   _____   1. I read an *entire* problem before I attempt a solution. (See Tip 1, page 51.)

_____   _____   2. I look up and try to understand anything in a problem that I am unfamiliar with. (See Tip 1, page 51.)

_____   _____   3. I determine the principle dealt with in a problem. (See Tip 2, page 51.)

_____   _____   4. I write down all of the information given. (See Tip 3, page 52.)

_____   _____   5. I determine what I am trying to solve for. (See Tip 4, page 52.)

_____   _____   6. I sit back and plan how I am going to solve a problem. (See Tip 5, page 53.)

_____   _____   7. I determine whether or not I need additional information in order to solve a problem. (See Tip 6, page 54.)

_____   _____   8. I determine if there is irrelevant information given in a problem. (See Tip 7, page 54.)

_____   _____   9. I solve problems in a neat and organized manner. (See Tip 8, page 55.)

_____   _____   10. I look for help if I am having difficulty doing a problem. (See Tip 9, page 56.)

_____   _____   11. I include all units when I am doing my calculations. (See Tip 10, page 57.)

_____   _____   12. I check my answers to see if they make sense. (See Tip 11, page 57.)

_____   _____   13. I pay attention to significant figures. (See Tip 12, page 57.)

_____   _____   14. I pay attention to stupid mistakes. (See Tip 13, page 58.)

_____   _____   15. I keep a list of important formulas and their significance to problem solving. (See Tip 14, page 58.)

_____   _____   16. I review any mathematical manipulations that I have difficulty with. (See Tip 14, page 58.)

Review the statements to which you have answered no. In addition, review the suggestions that you have already written to yourself on a separate sheet of paper. Then, do the following:

A. Prioritize the list of suggestions and select three behaviors that you can change that will help you to improve your problem-solving skills.

B. Write these behavior changes in the space below.

C. Be specific about what you want to change, how you are going to make the change and when the new behavior will be implemented.

D. See if these changes help you to better solve chemistry problems.

  1.

  2.

  3.

Do not try to make too many problem-solving changes at one time because this can be overwhelming and lead to frustration. Once you have implemented the above three changes and have seen improvement in your skills, we suggest that you again review areas in your problem solving that need improvement. Then go through steps A through D and watch for further improvement.

Remember, solving chemistry problems is not impossible. If you approach your problem solving systematically and follow a few simple guidelines, you will discover a method to solving chemistry problems.

You may make some false starts, but you learn just as much, if not more, in an incorrect attempt as you do if you do the problem flawlessly. Keep the following in mind:

1. You must be *actively* involved in the homework assignment and have paper, pencil, calculator, and 3 × 5 cards handy.

2. You must learn to *read* and *analyze* the problems.

3. You must learn to do the problems and keep track of every step in the solution.

4. You must learn to do the problems with *understanding*.

Analysis Sheet for Doing Chemistry Problems

You may want to use the following analysis sheet when you do chemistry problems in the future.

Check off each step as you do a chemistry problem.

____ 1. Read the entire problem and review the text, when necessary.

____ 2. Determine the problem topic.

____ 3. Write down all the information given.

____ 4. Determine what you are trying to solve for.

____ 5. Plan your method of attack.

____ 6. Determine the additional information that is needed.

____ 7. Determine what information is unimportant to solving the problem.

____ 8. Solve the problem in a neat and organized fashion.

____ 9. Include the units in your solution.

IF YOU ARE HAVING DIFFICULTY SOLVING A PROBLEM, DO THE FOLLOWING

____ 10. Reread the section in the text that covers a particular topic.

____ 11. Review the example problems in the text.

____ 12. Review the class notes.

____ 13. Consult your supplementary texts.

____ 14. Check your mathematics.

___   **15**. Discuss the problem with your study group.

___   **16**. Seek help from your professor or a tutor.

___   **17**. Look at the solutions manual—only as a last resort.

AFTER SOLVING THE PROBLEM, DO THE FOLLOWING

___   **18**. Check your answer.

___   **19**. Check your significant figures to make sure they are correct.

___   **20**. Check for stupid mistakes.

IT MAY BE IMPORTANT TO DO THE FOLLOWING FOR SOME PROBLEMS

___   **21**. Make a list of important formulas and how they are used.

___   **22**. Review mathematical manipulations if you are having difficulty.

6

TAKING CHEMISTRY TESTS

How can
you do well
on chemistry
exams?

?

Students often complain that chemistry tests are too difficult, too long, and that professors are too picky about little mistakes. There's no question that chemistry tests are challenging and that you can't fake your way through one.

Doing well on a chemistry test means that you have to know chemical formulas and equations, be able to apply specialized terms, be adept at analyzing and solving problems, and know how to do various mathematical manipulations. Having to possess a vast amount of information and a mastery of several problem-solving skills may sound overwhelming. However, it doesn't have to be, if you prepare for your exams correctly.

In addition to acquiring of knowledge and skills, you also have to be extremely careful about little mistakes. Losing points because of these mistakes is not simply a matter of your professor being picky. Students often interpret the accuracy of chemistry as pickiness. A "small" error in an equation, a formula, or units, can significantly change an answer to make it completely wrong. This would be very serious in a chemistry laboratory, and so your professor also makes it serious on an exam—which means that if you make one of these simple little errors, you could lose a number of points.

In this chapter we're going to focus on some skills that will help you better prepare for chemistry exams, and that you can use during exams to improve your performance. Remember that in addition to the tips in this chapter on test taking, the tips from our previous chapters all contain important information which will help you when preparing for your chemistry exams.

Let's begin with a look at some problems students sometimes have with chemistry tests.

Problems Students Have with Chemistry Tests

Read the following list of problem statements and place a check mark in the appropriate space to the left to indicate whether the statements are true or false for you.

YES   NO

_____   _____   1. I often feel overwhelmed by the vast amount of information I'm supposed to know for my chemistry exams.

_____   _____   2. I'm so nervous when studying for an exam, I'm unable to concentrate on what I'm studying.

_____   _____   3. I get so nervous the night before an exam that I have difficulty getting to sleep.

_____   _____   4. When I'm taking an exam, I often forget important information that I knew when I was studying.

_____   _____   5. I study hard, but my professor always seems to ask the wrong questions.

_____   _____   6. I always run out of time during chemistry exams.

_____   _____   7. I often make stupid mistakes on chemistry tests.

_____   _____   8. I'm easily distracted during chemistry tests, so that I get confused about what I'm doing.

_____   _____   9. I often make mistakes on chemistry exams because I misread the questions.

_____   _____   10. I think I'd do a lot better on my chemistry exams if I wasn't so nervous about them.

_____   _____   11. I don't really think I'm that smart, and that's why I don't do better.

If you marked yes to any of the above statements, you could benefit from improving your test-taking skills. Read the following tips for improving your test-taking skills.

If none of the above statements were true for you and if you don't think you need to improve your performance on chemistry tests, skip ahead to the next chapter, "The Chemistry Laboratory."

Tips for Improving Your Chemistry Test-Taking Skills

As you read the tips for improving your test-taking skills, keep in mind the problems you experience with chemistry tests. Look for ways to change your behavior before, during, and after exams in order to improve your performance on tests and thus improve your grades.

When reading the following, note on a separate sheet of paper anything that you want to do differently to improve your test-taking skills.

BEFORE THE EXAM

1. *Prepare early.* Start to prepare for your chemistry exams on the first day of class by

   a. collecting and organizing materials for the course.

   b. finding out the dates of exams.

   c. scheduling regular study times throughout the semester.

2. *Schedule several study sessions for the exam.* Your study time for chemistry exams should be spread over several sessions so that you have a number of exposures to the material. Research has shown that you learn more and your retention is better if you learn material over a number of shorter sessions rather than one very lengthy session. This is the reason why cramming for two days before a test doesn't work very well. Your mind needs time in between study periods to absorb the new material.

   If you are in a situation where you need to cram for a test, your best bet is to try to pick up the main points and then spend time reciting and doing problems. Once again, recitation is the key to remembering.

3. *Know the exam details.*

   a. Before the exam find out the following:

   - What types of questions will be asked (i.e., essay, problem solving, short answer, multiple choice, etc.)?

   - Will partial credit be given on problems?

   - Will you be penalized for guessing, especially on multiple choice?

   - How much time will you have to take the exam?

   - How much will the exam count toward your final grade?

- What percentage of the exam will be taken from lecture, from the textbook, and from outside readings?

- Will there be a special seating arrangement?

  b. If your professor hasn't given you the information you need to answer the above questions, ask him or her.

4. *Look at copies of tests from previous years if they are available.* These are excellent study aids, but do not study only the material covered on these old exams, as professors often make changes on exams. Use the old exams to give you an idea about the professor's exam format and to help you gain some insight into how the professor thinks.

If you are unable to get copies of previous exams, ask students who have taken the course before about the examination format. Again, be cautious. Your professor may have made some changes that will surprise you!

5. *Develop "summary sheets" as preparation for your test.* The book *How to Study In College*, by W. Pauk recommends the use of summary sheets as a good way to review for exams. Summary sheets are consolidations of your lecture notes and of your textbook notes which are used for recitation. Pauk says these are helpful for review because

   a. In making up the summary sheets, you review the notes you took throughout the semester.

   b. The information in your summary sheets is categorized according to topics, so that it's better organized and easier to recall during exams.

   c. Your condensed set of notes can be used for quick review before your exam.

   d. You can't remember everything. Summary sheets stick to the main points and the most essential information.

6. *Work one or two example problems from each principle that's been discussed.*

   a. For most students, it would be impossible to rework all the assigned problems when studying for a chemistry exam. One good way to check and see if you are prepared for the exam is to work two of the assigned problems for each concept. You can quickly check your ability to do the problems from the solution sheets you saved. (See "Chemistry Problem Solving," Chapter 5.)

b. If you have difficulties, work additional problems, and make sure you understand the reason(s) for your difficulties. For example, if you don't understand the significance of a mole in stoichiometric calculations, you will have difficulty doing this type of problem.

7. *Use your 3 × 5 cards to test yourself for review.* (See Tip 11 in Chapter 4 for information on how to make up these cards.) Test yourself on your knowledge of formulas, definitions, and terms using these cards. If there is one formula that you just can't remember, keep that particular card handy and spend time going over it and testing yourself until you feel confident about it. Continue going over the cards which are difficult for you.

8. *Try studying in the same room where you'll be taking your chemistry exam.* This will allow you to develop cues associated with the exam room which will help you to remember the material you study. This

procedure will facilitate the recall of information you studied while in the exam room and should also help reduce test anxiety.

9. *Learn how to control any anxiety you experience while studying for your exam.* Many students worry so much about how they're going to perform on exams that they have difficulties studying because their worries interfere with their concentration. If anxiety interferes with your ability to study, there are several things you can do to control it:

   a. *Learn how to relax your body.* Information about the Benson meditation technique and on progressive relaxation are included in Appendix A. Both are easy to do and will help you relax physically. Just learning to take in a few deep breaths, slowly exhaling as you count to three, will be helpful.

   A regular form of cardiovascular exercise, such as jogging or swimming, is also helpful for physical relaxation.

   b. *Learn how to relax your mind.* Your mind cannot think two different thoughts at once. Thus, if you're thinking about your worries, you will not be able to concentrate on chemistry. At the same time, if your full attention is on your chemistry, you will not be able to concentrate on worrying.

   Make a mark on a piece of paper every time your mind begins to wander to worries about how you're going to do on the exam. Remember that monitoring undesirable behavior will help it decrease in frequency.

   Recognize when you start worrying. Have a word that you can say to yourself, such as "focus," as a signal you can use to stop worrying and to get back into studying.

   c. *If none of these techniques works and you still experience a lot of anxiety while studying,* seek help from a counselor or psychotherapist. Keep in mind that controlling anxiety is not as easy as it may seem. You are trying to break a habit that you have been practicing for years.

10. *Use relaxation and creative visualization for 20 minutes a few evenings per week beginning several weeks before the exam.*

    a. Use the relaxation exercise mentioned in Tip 9 and also try the following five to ten minute creative visualization exercise:

Begin by taking in a few deep satisfying breaths, just letting yourself relax. Then try to picture the room where you will be taking the chemistry exam. Picture in detail what it looks like, smells like, feels like. Picture yourself sitting and taking the exam. First picture yourself being nervous, what your body looks like when you're nervous, what your facial expression is like. Try to feel the nervousness. Now erase that picture as you take in a few deep breaths, slowly exhaling, counting to three, telling yourself to relax.

Next, picture yourself coping with your nervousness by seeing yourself taking in a deep breath and slowly exhaling, changing your body position, focusing your attention on the test, tuning out the rest of the world. Picture yourself positively, confidently, concentrating on the exam, in a body position that you think might be typical of you when working hard on a test and doing a good job concentrating. Be as clear and specific with your imagery as you can. Think of how you might look and feel if you were doing well on the exam. Try to think of yourself as being clear and focused, challenged by the exam, involved with it, and doing well. Do not criticize yourself while doing this exercise.

11. *Do the following the night before the exam*:

   a. *Relax*. You can use the relaxation and creative visualization exercises discussed in Tips 10 and 11, if you've already practiced them. If you try the exercises for the first time the evening before the exam and aren't able to relax or use the visualization right away, you may end up getting frustrated, which will increase your anxiety. You don't need to be doing new things that increase your anxiety the night before your test.

   b. *Review your notes.*

   c. *Get your materials ready*. Set out your pencils, calculator, and anything else you need for the exam.

   d. *Get a good night's sleep*. The night before the exam is not a good night to party or go to a movie, even if you're finished with your studying and feel confident about the material. The evening before the exam is the time for a relaxed review without a lot of extraneous thoughts interfering with your concentration on chemistry.

DURING THE EXAM

12. *Come to the exam early and come prepared.*

    a. Arrive early to the exam. Find a good place to sit where you have some privacy and can see the chalkboard. It is usually not a good idea to sit with a friend as friends are often great distractors.

    b. You've probably noticed upon entering the room where a test is to be given, that students are often more noisy than usual. The reason for this is that they're nervous about the exam. It's more calming to sit by yourself, not to talk with anyone, and to quietly review your notes. This will allow you to get your mind focused on the exam and you won't have a lot of extraneous thoughts interfering with your recall of pertinent information.

c. Come prepared. Have two sharpened pencils and a working calculator that you know how to use.

13. *As soon as you get the exam, jot down any important definitions or formulas* that you know you'll be using on the exam. It is helpful to write down this information before your thoughts are interrupted by attending to the exam questions.

14. *Briefly glance over the entire exam* before working on it, to get an idea of what the questions are, which are worth the most points, and which ones will be the most difficult.

15. *Read the directions very carefully.*

   a. Too often students lose points because they do not follow a few simple directions. For example, most chemistry professors will not give any credit for an answer to a problem unless a solution is included—even if the answer is correct. The professor has usually stated this in the directions—so read or listen to them carefully.

   b. If you don't understand the directions, ask your professor.

16. *Make corrections on your exam.* Sometimes there are typographical errors on exams. The professor will usually announce this at the beginning of the exam and also write the corrections on the board. Make these corrections on your exam immediately because you may forget to do so after you become involved with the exam. Students often lose exam points because they fail to make these corrections.

17. *Read a question completely and carefully before you start to work the problem.*

   a. You must clearly understand what is being asked in a question. Sometimes students solve for the wrong answer because they jump to conclusions about a problem or fail to read the last sentence.

   b. If, after reading a question carefully, it seems unclear to you, ask your professor about it. Sometimes there are questions or directions that are ambiguous, or mistakes that the professor missed while proofreading the exam.

18. *Start with the problems that are easiest for you.* This will help you build confidence and ease your anxiety. This does not mean that you should skip difficult problems altogether.

19. *Organize your solution to a problem*. Remember that professors have to grade a lot of exams and they get irritated and sometimes don't even grade a question if they have to spend time searching for the solution.

Present your solutions to problems in a neat and organized way that is easy to follow. Use scrap paper (if permitted) or use the backs of the exam pages when you are solving a problem that you are unsure of. Rewrite the final solution on your exam paper.

20. *Budget your time during the test*. Have a watch or clock to pace yourself. Try to get the most points in the time you have available. If you're running behind time, don't panic. Prioritize and work the problems that are worth the most points.

21. *Work multiple choice problems just as you do other problems on a chemistry exam*. Often students get used to answering multiple choice questions by just thinking about them. This doesn't work well on chemistry exams because most multiple choice questions will require you to do some calculations in order to get the correct answer.

22. *Don't leave the exam early*.

    a. And don't be intimidated by those students who do leave early. Exam scores do not reflect the amount of time taken to do the exam. Some students leave early because they do not know the material.

    b. Remember that on chemistry exams, it's easy to make simple errors that can cost you a lot of points. Even if you have already checked for mistakes, use your extra time to check again. (See Tip 23 for more information on what to check.)

23. *Allow time to check over your exam*.

    a. When you finish your exam, check it over for the following:

    ■ Have you skipped any problems? This is especially easy to do on multiple choice questions.

    ■ Have you paid attention to significant figures in your answers? Recheck every answer and express it to the correct number of significant figures.

    ■ Have you included the units on your answers?

    ■ Does your answer make sense? Remember, you cannot have 1.08×10^{-10} atoms in a mole.

■ Have you finished each problem or did you stop in midsolution?

b. If time permits, check the following:

■ Go over your calculations to make sure you haven't made a calculation error.

■ Recalculate as many problems as possible and see if you get the same answer you did the first time. Do this on scrap paper and redo the entire solution without looking at your original solution. If you get the same answer twice, it's more likely to be the correct answer.

24. *Manage your anxiety during the test.* Perhaps you're a student who experiences a lot of anxiety while taking chemistry exams.

a. Your palms may sweat, your heart may pound, or your mind may simply go blank. Sometimes your worries about these symptoms make it difficult for you to concentrate on the exam.

b. You also may be spending time catastrophizing. Catastrophizing works something like this: You have difficulty with one of the problems, and right away your mind jumps to "I'm going to bomb the exam." It then jumps to "I'm going to fail chemistry and all my other courses too." You then picture one or both of your parents looking very angry or disappointed, or you picture yourself flunking out of college and not being able to get a job, and of course this means that you're stupid, just as you've always known, but now everyone else will know too and that will be awful, etc., etc.

c. Most students are good at catastrophizing. The problem is, that catastrophizing increases anxiety. It's easier to concentrate on developing your catastrophic scenario than it is to concentrate on your chemistry exam. What can you do about it? Remember some of the points we made in Tip 10:

You first need to learn to relax your body. Take a few deep breaths, exhaling slowly.

You need to relax your mind and bring your focus back to your chemistry exam. When you start worrying about your heart beating fast or worrying about what your parents will say, tell yourself to STOP IT, that you need to focus on the exam and get involved with what you're doing. It may help to picture a smiling coach sitting on your shoulder cheering you on, reminding you to focus on the exam.

AFTER THE EXAM

25. *Go over your exam carefully as soon as you get it back.*

    a. Determine exactly why you made the mistakes you did. Students' mistakes on chemistry exams are often due to the following:

    - Insufficient studying for the exam and not understanding the material.

    - Incorrect use of significant figures.

    - Not following directions.

    - Not reading the entire problem and, thereby, solving for the wrong answer.

    - Not making corrections that were written on the board.

    - Using mathematical manipulations incorrectly (for example, multiplying instead of dividing by a constant).

    - Presenting a solution to a problem in a form that is disorganized and difficult to read.

    - Having a calculator fail in the middle of the exam.

    - Solving a problem both incorrectly and correctly without specifying to the professor which solution is the correct one.

    - Illegible handwriting.

    - Not including units in the final answer.

    - Spending too much time on one problem at the expense of solving other problems.

    - Not completing a problem.

    - Coming in late for the exam.

    - Inability to recall information due to nervousness.

    - Inability to concentrate on the exam due to nervousness.

    b. After you determine the source of your errors, write down what you need to do differently.

    c. If you are unsure about where you went wrong, bring your exam to your study group, a tutor, or your professor and ask them for help so that you can improve your test scores and do well on that next chemistry exam.

26. See "Test-Taking Distraction and Remedy Check List," Appendix B.

Test-Taking Check List

To help you clarify what changes you need to make to improve your chemistry test-taking skills, answer yes or no to the following statements. If necessary, review the appropriate tips for "no" statements.

YES   NO

_____ _____ 1. I begin preparation for my chemistry exams as soon as the semester starts. (See Tip 1, page 73.)

_____ _____ 2. I devote several study sessions to preparation for a chemistry exam. (See Tip 2, page 73.)

_____ _____ 3. I anticipate exam questions and format prior to the exam. (See Tip 3, page 73.)

_____ _____ 4. When available, I look at copies of tests from previous years. (See Tip 4, page 74.)

_____ _____ 5. I use summary sheets for exam preparation. (See Tip 5, page 74.)

_____ _____ 6. I am able to work one or two problems for each concept before taking an exam. (See Tip 6, page 74.)

_____ _____ 7. I use the 3 × 5 card technique to test myself. (See Tip 7, page 75.)

_____ _____ 8. I do some of my exam preparation in the room where my exam is given. (See Tip 8, page 75.)

_____ _____ 9. Anxiety *does not* interfere with my exam preparation. (See Tip 9, page 76.)

_____ _____ 10. I practice relaxation and creative visualization on a regular basis. (See Tip 10, page 76.)

_____ _____ 11. I am able to relax the evening before a chemistry exam. (See Tip 11, page 77.)

_____ _____ 12. I come to chemistry exams early, find a good seat where I can see the board, and relax. (See Tip 12, page 78.)

_____ _____ 13. As soon as I am given a chemistry exam, I jot down important definitions and formulas on the exam paper. (See Tip 13, page 79.)

_____ _____ 14. I briefly glance over a chemistry test before I begin work on it. (See Tip 14, page 79.)

_____ _____ 15. I carefully read the directions for the exam and for each problem. (See Tip 15, page 79.)

_____ _____ 16. If I don't understand the directions, I ask the professor for help. (See Tip 15, page 79.)

_____ _____ 17. I listen carefully to the professor's announcements about the exam and immediately make any corrections. (See Tip 16, page 79.)

_____ _____ 18. Before I work an exam problem or answer a question, I read it carefully. (See Tip 17, page 79.)

_____ _____ 19. I begin a chemistry exam with the problems or questions that are easiest for me. (See Tip 18, page 79.)

_____ _____ 20. My answers to exam questions are neat and organized and easy for my professor to read. (See Tip 19, page 80.)

_____ _____ 21. During a chemistry test I budget my time. (See Tip 20, page 80.)

_____ _____ 22. I solve the problems on multiple choice questions, as opposed to just thinking about an answer. (See Tip 21, page 80.)

_____ _____ 23. After I finish a chemistry exam, I spend any remaining time checking my answers. (See Tips 22 and 23, page 80)

_____ _____ 24. Anxiety *does not* interfere with my chemistry exam performance. (See Tip 24, page 81.)

_____ _____ 25. When my graded exam is returned, I know how to go over it to prepare for my next exam. (See Tip 25, page 82.)

_____ _____ 26. I use the "Test-Taking Distraction and Remedy Check List." (Appendix B, page 100.)

Review the statements to which you answered no. In addition, review the suggestions for improving your chemistry test-taking skills which you have already written on a separate sheet of paper. Then, do the following:

A. Prioritize these lists and select three behaviors that you can change that will help you to improve your test-taking in chemistry.

B. Write these behavior changes in the space below.

C. Be specific about what you want to change, how you are going to make the change, and when the new behavior will be implemented.

D. See if these changes help you take the exams in chemistry.

    1.

    2.

    3.

Do not try to make too many changes at one time, because this can be overwhelming and lead to frustration. Once you have implemented the above three changes in your chemistry test-taking behavior and have seen improvement in your skills, we suggest that you again review your chemistry test-taking habits. Then go through steps A through D and watch for further improvement.

7 THE CHEMISTRY LABORATORY

The chemistry laboratory is a very important part of the chemistry course for many reasons.

1. The laboratory allows you to become familiar with some of the theories discussed in the lecture.

2. The laboratory is the only way for you to learn the techniques that are so important not only in chemical research but in most scientific laboratories.

3. You will discover that doing quality work in the laboratory requires a great deal of patience and care.

4. The laboratory will allow you to experience some of the joys and frustrations associated with being a research scientist, and that's exciting.

In this chapter, we will examine what you can do to improve your performance in the laboratory and to make the laboratory more enjoyable for you.

Let's take a look at some problems chemistry students often experience in the laboratory.

Problems Students Have in the Chemistry Laboratory

Read the following list of problem statements and place a check mark in the appropriate space to the left to indicate whether the statements are true for you.

YES   **NO**

1. I don't understand the experiment even after I read the laboratory manual.

2. I sometimes can't do the prelaboratory assignment.

3. I sometimes don't understand the prelaboratory lecture.

4. I don't always know how to use the equipment.

5. I feel stupid if I have to ask the professor questions.

6. I am afraid of working in the laboratory.

_____ _____   7. I am often confused in the laboratory.

_____ _____   8. I feel clumsy in the laboratory.

_____ _____   9. I often can't do the required calculations for the experiment.

_____ _____   10. I never seem to understand the point to the experiment.

_____ _____   11. I feel rushed when trying to finish experiments.

_____ _____   12. I spend a lot of time bumbling around trying to start the experiment.

If you marked any of the above statements yes, you could benefit from improving your laboratory skills. Read the following tips for improving your laboratory skills.

If none of the above statements were true for you and you don't experience any problems in the chemistry laboratory, congratulations. You are finished with the book.

Tips for Improving Your Laboratory Skills

As you read the tips for improving your laboratory skills, keep in mind the problems you experience in the laboratory. Look for ways to change your laboratory behavior in order to improve your skills.

As you read the following tips, write down on a sheet of paper anything that you want to do differently to improve your laboratory skills.

1. Read the laboratory experiment before coming to the laboratory. You may say to yourself, "I do read the laboratory manual but I still don't understand the experiment." Keep the following in mind:

   a. Reading the experiment will probably not give you a thorough understanding of the entire procedure and theory behind it. However, it will allow you to become vaguely familiar with the experiment.

   b. During your readings, you will look at the diagrams of the apparatus to be used, read the discussion on the theory, and do some sample calculations.

   c. The instructor's prelaboratory discussion will clarify the experimental procedure. (Keep in mind that no one expects you to master everything before the laboratory starts!)

2. *Do the prelaboratory assignment before your laboratory*. This will help you to better understand the procedure and calculations that are associated with a particular experiment.

3. *Write down any questions* that may arise during your reading of the experiment or while you are doing the prelaboratory assignment. If there aren't too many questions, save them until you go to the laboratory. If your questions aren't answered during the prelaboratory lecture, ask your instructor before you start the laboratory. If there are a lot of questions, get help either from members of your study group, a tutor, or visit the instructor during office hours.

4. *Get to the laboratory early* and be ready to listen to the instructor when he or she starts. Arriving late for laboratory not only causes you to miss the prelaboratory lecture and, thereby, any important information about the laboratory, but it also serves to distract those students who are listening to the instructor. (It's also likely to bother your instructor.)

5. *Listen attentively to the prelaboratory lecture*. The prelaboratory lecture is designed to bring out the important points of the laboratory and, hopefully, clear up any questions students may have. There is usually a discussion of the theory behind the experiment, a sample calculation, and a demonstration of the experimental procedure.

6. *Ask questions* before you start the experiment if anything is unclear about the procedure. Remember, there are potentially dangerous situations that arise during the performance of any laboratory experiment.

7. *Ask questions* during the performance of the experiment either about the procedure or the theory upon which the experiment is based. This will help you better understand the experiment and the lecture material.

8. *Monitor the experiment as it progresses*. Do a quick calculation to see if the results are reasonable. This will tell you if you are doing the experiment correctly, whether you need to repeat the experiment, or if you need to change the experimental conditions.

9. *Wear your safety goggles at all times*. Even if you are not doing the experiment and are simply chatting with another student or doing some calculations, the fact that other students are working creates a situation which could be dangerous for you. For example, another student

could break a beaker and a piece of the broken glass could fly into your eye.

10. *Follow all safety rules.* Treat all chemicals with respect and wipe up any spills. Report all accidents to the instructor. Keep your work area clean.

11. *Record all your data* (neatly, in ink and with units) in the laboratory notebook, not on scrap paper or paper towels which can be lost easily. In addition, you can make mistakes when you are copying your original data onto the data sheet. If you throw away or misplace your original data you are up a creek without a paddle.

12. *Do not erase your original data.* Simply put a line through it if you make a mistake. It is possible that these data are correct, and you may want to use them at a later time.

13. *If you make a mistake, repeat that part of the experiment.* Everyone makes mistakes, but good scientists repeat an experiment when a mistake has been made. Some of the benefits of repeating experiments are:

  a. You are forced to think about the possible sources of error. This results in a better understanding of the procedure and the theory upon which the experiment is based.

  b. If you develop good laboratory techniques during your introductory course, you will carry these over to all chemistry and science courses you take in the future.

14. Do your calculations as soon as possible after the laboratory is finished. If time permits, do them while you are in the laboratory.

  a. This will allow you to take advantage of the fact that the instructor is available to answer any questions that may arise.

  b. If you procrastinate and wait until the night before your next laboratory session, you will not only have to prepare for the new laboratory but review the previous laboratory.

Laboratory Skills Check List

To help you clarify what laboratory behavior changes you need to make, answer yes or no to the following statements. If necessary, review the appropriate tips for the "no" statements.

YES   NO

_____ _____ 1. I read the laboratory experiment before coming to the laboratory. (See Tip 1, page 88.)

_____ _____ 2. I complete the prelaboratory assignment before coming to the laboratory. (See Tip 2, page 89.)

_____ _____ 3. I write down any questions that arise while I am reading the laboratory experiment or working on the prelaboratory assignment. (See Tip 3, page 89.)

_____ _____ 4. I try to get my questions answered before the laboratory starts or at the beginning of the laboratory. (See Tip 3, page 89.)

_____ _____ 5. I get to the laboratory early enough so that I am ready when the laboratory begins. (See Tip 4, page 89.)

_____ _____ 6. I listen attentively to the prelaboratory discussion. (See Tip 5, page 89.)

_____ _____ 7. I ask questions during the laboratory and don't let things slide by. (See Tips 6 and 7, page 89.)

_____ _____ 8. I take the time to think about the laboratory experiment while I am performing it and do not go through it blindly. (See Tip 7, page 89.)

_____ _____ 9. I pay attention to the experimental results as the laboratory progresses. (See Tip 8, page 89.)

_____ _____ 10. I wear my safety goggles whenever I am in the laboratory. (See Tip 9, page 89.)

_____ _____ 11. I observe all safety rules. (See Tip 10, page 90.)

_____ _____ 12. I record all my data on the sheet and not on scrap paper. (See Tip 11, page 90.)

_____ _____ 13. I do not erase my mistakes on the laboratory data sheet but instead cross them out. (See Tip 12, page 90.)

_____   _____   **14.** I repeat any part of the experiment when I obtain results that seem inconsistent or incorrect. (See Tip 13, page 90.)

_____   _____   **15.** I do my calculations immediately after the laboratory and don't wait until the night before the next laboratory. (See Tip 14, page 90.)

Review the statements to which you have answered no. In addition, review the laboratory skills improvement suggestions that you have already written to yourself on a separate sheet of paper. Then, do the following:

A. Prioritize the list of suggestions and select three behaviors that you can change that will help you to improve your laboratory skills.

B. Write these behavior changes in the space below.

C. Be specific about what you want to change, how you are going to make the change, and when the new behavior will be implemented.

D. See if these changes help you to improve your laboratory skills.

  1.

  2.

  3.

Do not try to make too many laboratory behavior changes at one time because this can be overwhelming and lead to frustration. Once you have implemented the above three changes and have seen improvement in your skills, we suggest that you again review areas in your laboratory skills that need improvement. Then go through steps A through D and watch for further improvement.

Remember, the laboratory can be an exciting learning experience. It's a place for you to discover what being a laboratory scientist is all about. If you prepare for the laboratory and follow a few simple guidelines, you should have little difficulty and you could find the experience to be quite interesting and rewarding. The key words to remember when working in the laboratory are

1. Prepare.

2. Listen and follow instructions.

3. Be careful.

4. Focus on what you are doing.

5. Enjoy yourself.

Congratulations! This is the end of the book. Go back and review any chapter where you are still having difficulties. Then make the recommended changes and *practice* them.

APPENDIX A

The Relaxation Response (Meditation)

I. *Definition*:   A technique, not necessarily part of a religious or philosophic system, to help you relax and turn your energy inward. Researchers have identified a particular "relaxation response" that accompanies meditation.   Characteristics of this response include a lowered heart rate, decreased muscle tension, lowered blood pressure, lowered lactic acid levels, and increased alpha brain wave activity. The technique is easily learned and has many potential benefits.   It originated in Eastern cultures and has many very useful applications for Western cultures where there is little emphasis on encouraging people to respond to their need for inner tranquility.

II. *Meditation and Anxiety*: The regular practice of meditation can decrease anxiety in several ways. The physical relaxation achieved during meditation is in direct opposition to the physical tension accompanying anxiety; consequently, relaxation counteracts anxiety.   The regular practice of meditation tends to build a regular slowing down cycle into your general pattern of activity. The effects of the relaxation response are cumulative, and in time you will experience a general decrease in tension. The mental relaxation and turning inward of your mind provides a regular mental respite from the trials and tribulations of everyday life. It helps you to break the thinking/worrying cycle that so often causes anxiety and keeps it going. The side benefits of meditating and being more relaxed are important, too.   When you are relaxed you tend to have energy, to be more effective interpersonally, to have greater self-confidence, and to use fewer drugs. This, in turn, gives you a better general feeling of competence and self-confidence.

Adapted from J. Bryer and J. Archer, *Anxiety Management Sessions*, University of Delaware.

III. *The Process*: Meditation is a natural, simple process. It is easy to learn and to practice. There are several important things to remember.

1. Start by closing your eyes and trying to deeply relax all the muscles of your body as best you can.

2. Pick a pleasant sounding word or phrase (for example, "relax," "calm," "serene," "one") and repeat it over and over again to yourself each time you exhale. The purpose of this is to help you shift from logical, externally oriented thought to a more inward, relaxed consciousness. As you relax you'll notice that the repetition of the word or phrase will become quieter and quieter.

3. Maintain a passive attitude. Interfering thoughts and feelings will come. Don't fight them, let them come and then gently go back to repeating your word or phrase. Don't worry about your performance. The process of meditation develops naturally.

4. Use the same comfortable position whenever you meditate. Pick a position that allows you to relax and requires little work from your muscles. Don't pick a position in which you normally sleep.

5. Meditate in a quiet environment. Try to be alone in a room with the door shut and where you are not likely to be disturbed.

6. Meditate twice a day for 15–20 minutes. It is usually best to pick a regular set of times, say before breakfast and before dinner.

7. When you are ready to end your meditation, don't stop too quickly. Leave your eyes closed 20 or 30 seconds as you come out.

Self-Relaxation Program

STAGE I   Lie on your back in a darkened room, with eyes closed, arms at your sides, legs uncrossed. Place pillows under your neck, knees, and feet, if that increases your comfort. Loosen or remove any binding clothing (shoes, belts, tight collars) as these can contribute to muscle tension.

Taken from M. Weissberg and F. Breme, *Test Taking and Test Anxiety*, University of Georgia Counseling and Test Center, 1984.

Spend 30–45 minutes alternately contracting and releasing parts of your body, one at a time. When you are tensing, try to make your muscles tighter and tighter and hold for about 10–20 seconds. Then let go *very slowly* and *feel* the relaxation developing. When you think you have reached your limit, keep telling yourself to "let go further and further." This will help you sense how real relaxation feels. The exercises should follow a logical order through your body. Here is the recommended sequence:

1. *Hands and Arms*

    ▪ Clench (then relax) each fist.

    ▪ Bend both elbows and flex your biceps, hard.

    ▪ Rigidly straighten both arms.

2. *Face, Head, and Neck*

    ▪ Wrinkle your forehead until you feel tension moving across your scalp.

    ▪ Frown deeply, to tense every muscle in your face.

    ▪ Close your eyelids as tightly as possible.

    ▪ Clench your teeth and notice the tension in your jaws, cheeks, throat, and neck.

    ▪ Press your tongue against the roof of your mouth.

    ▪ Press the back of your head firmly into your pillow or against the floor.

    ▪ Push your chin against your chest and strain your head forward.

    ▪ Shrug your shoulders up to your ears.

Pause here. *Think* about a feeling of relaxation spreading from your fingertips up to your scalp and down to your neck and shoulders.

3. *Midsection*

    ▪ Inhale as deeply as you can and hold before *slowly* exhaling. Repeat.

    ▪ Tense your abdominal muscles as tightly as possible *without* pulling them in.

    ▪ Pull in your stomach and hold.

- Arch your back and hold as long as you can.

Pause again. Think about relaxation gradually spreading through all the muscles of your chest, back, and stomach. Then imagine that feeling spreading even deeper into your head and shoulders and down your arms.

4. *Legs and Feet*

- Press both heels hard against the floor.
- Point your feet and toes away from your head as far as possible.
- Point your toes toward your head.

Pause and think about feeling relaxed, starting with your toes and slowly moving up your entire body. Try to imagine the tension you'd feel if you were actually lifting your legs. Enjoy your relaxed state as long as you like, then stretch, yawn, and get up. If you feel yourself becoming tense later on, try to recapture the "letting go" feelings you experienced.

Repeat the exercises daily for two weeks and try not to become discouraged. For most people, the "breakthrough" doesn't occur until about midway through Stage II.

Stage II focuses on mental methods that can help peel away layers of tension and anxiety. In short, this stage produces relaxation on demand.

STAGE II   Spend about 30 minutes a day following this schedule:

1. Lie comfortably in a darkened room, breathe deeply several times, and feel a warm, heavy, relaxed sensation spread through your body as you slowly exhale. (You should know by now what a "relaxed" feeling is.) Think about a wave of calm flowing over your body in a slow, logical sequence.

2. Spend a minute or so thinking the word "relax" each time you exhale. Next, think of other stimulus words: calm, serene, tranquil, warm, confident, restful, peaceful, and the like. Pause after each word and try to associate feelings of relaxation with it. Pick two words that seem most calming to you and repeat them slowly about 20 times while you relax more and more deeply. With practice, you'll be able just to say those words to yourself in tense situations and touch off relaxed feelings.

3. For about 15 minutes, imagine relaxation spreading slowly through your body, starting at your forehead and ending with your toes. Think of the sensations involved as it spreads from one muscle or area to another.

4. Imagine a blank chalkboard, then put the numbers 1 through 10 on it, one at a time. As each appears, try to relax more deeply and capture the sensations suggested by a different stimulus word (warmth, tranquillity, serenity). By the time you reach 10 you should be totally relaxed.

Stage III is spent on conjuring up relaxed feelings while sitting, standing and walking, as well as while lying in a darkened room. The goal is to enable you to relax anywhere in any situation. It's a lot like driving a car. When you're first learning, you're very conscious of everything you do. But, with practice, it all becomes automatic.

STAGE III   Start each daily session lying comfortably on your back, breathing deeply several times and thinking about relaxation spreading through your body. Use your stimulus words and try to capture feelings of calmness within yourself. After you feel tension begin to drain away, move on to the next steps:

1. Imagine a scene you find very relaxing: a sunny beach, a cabin in the woods, a snowy evening, a mountain lake, a fireside. Savor the scene for about five minutes. This is your "personal relaxing image" and, like your key stimulus words, can be used to diffuse tense situations as you become adept at reacting to it.

2. Sit comfortably in a chair, with eyes closed, arms at your sides. One at a time and then together, raise your arms. Feel the tension of holding them up, then let them flop down in release. Repeat, holding a deep breath as you raise your arms, then exhaling as they flop. While your arms are up, concentrate on keeping the rest of your body relaxed.

3. Stand up and try to recapture feelings of relaxation, especially in your shoulders, stomach, and arms. With your eyes still closed, walk back and forth a short distance, swinging your arms gently and working to switch off any tension that may be creeping in.

4. Breathe slowly and regularly while standing still and thinking about relaxation spreading slowly through your body. Use your stimulus words to deepen the feeling.

5. Lie down and see how quickly you can regain a calm sensation all over your body. Then enjoy your personal relaxation image for at least one minute.

Like other students of relaxation, you may find one method more helpful than the others. In that case, use it. Daily practice is the best way to learn which parts of the program are most helpful in making you feel better or respond more calmly to pressure. After a while, your body's own rhythm will tell you when you need to use the techniques.

APPENDIX B

Test-Taking

Distraction and Remedy Check List

Distractions are the most prevalent problem when a person takes a test. When one is overly distracted by nervousness, sleepiness, or whatever, it becomes difficult to concentrate on the test itself. Many distractions are downright debilitating and must be controlled. Below is a list of things you can do to help avoid distractions. Use this list before every test you take.

DISTRACTIONS	REMEDY
Sleepiness	Get a good night's sleep before the test.
Nervousness/anxiety	Study well ahead of time and don't cram at the last minute. Cramming leads to panic and insecurity.
Full bladder	Go to the bathroom right before the test, not back at the dorm or apartment.
Upset stomach or tense stomach muscles	Relax stomach muscles and take an antacid, if necessary, before and/or during the test.
Tense neck and shoulder muscles, tightness in chest	Roll head clockwise and then counterclockwise in a full circle (2–3 times each direction).
	Roll shoulders in a circle 2–3 times in one direction and then 2–3 times in the other direction.

	Do five deep-breathing exercises (inhale all the air you can and then *slowly* exhale). Let go of half your tension with each exhale.
Not having what you need	Bring pencils, pens, calculators, blue books or whatever is needed so you don't panic when you discover you forgot something.
Tightness around neck, chest, waist, middle, and feet.	Wear loose fitting and comfortable clothes.
Handsome man or beautiful woman	Sit where you can't see, hear, or smell them.
Windows	Sit away from windows.
Concern about time	Wear a watch and *use it* to pace yourself.
Panic because you don't know the answer.	Do easy questions first. Memory dump. (Jot things down as you think of them, even when doing other questions. Organize answer later.)

Remember that it is your task to do well on all tests. These remedies can increase your chances of doing just that, so don't be ashamed or embarrassed to do these things. *You* are the one who needs good grades, so do what you have to do to get them. Don't worry about what someone else thinks.

From M. Weissberg and F. Breme, *Test Taking and Test Anxiety,* University of Georgia Counseling Center, 1984.

REFERENCES

Ellis, D. B. *Becoming a Master Student*. College Survival Inc., Rapid City, South Dakota, 1984.

Langan, J. *Reading and Study Skills*. McGraw-Hill Book Company, New York, 1978.

Learning Skills Program. University of Utah Counseling Center, 1984.

Paraprofessional Study Skills Materials. Northern Michigan University Counseling Center, 1986.

Pauk, W. *How to Study in College*. Houghton Mifflin Co., Boston, 1984.

Weissberg, M. and F. Breme. *Test Taking and Test Anxiety*. Counseling and Testing Center, University of Georgia, 1984.

PLEASE COMPLETE THIS FORM AFTER YOU HAVE FINISHED THE BOOK.

We would appreciate faculty, staff, and student evaluations of our book.

Completed evaluation forms can be sent to Margaret MacDevitt, 423 East Hewitt, Marquette, MI 49855.

1. Please check one of the following: Student _____ Faculty _____ Staff_____

Students, Faculty and Staff: Indicate your answers to Questions 2 and 3 by circling the whole number along each scale that best expresses your feelings.

Faculty and Staff, please respond to Questions 2 and 3 according to how useful you think this book would be *for students*.

2. How useful was this book to you?

    Of no use at all   1 2 3 4 5 6 7   Very useful

3. After reading this book, how do you expect your chemistry study skills behavior to change?

    Much less effective behavior   1 2 3 4 5 6 7   Much more effective behavior

4. Suggestions for improving the book:

_____

_____

_____

_____

_____

5. Comments:

_____

_____

_____

_____